"*Learning to Teach Science in the Secondary School* (5th Edition) is a useful guide providing invaluable advice for trainees and early career Science teachers alike. The new chapters on using research equip teachers for the research informed landscape that is shaping modern Science education. A superb addition to any trainee Science teacher's reading list."

Holly Kirkbride, *Director of Science, Ted Wragg Trust*

"An authoritative guide, written by experts, which will be important reading for all those involved in science education."

Mark Winterbottom, *Professor of Science Education, University of Cambridge*

LEARNING TO TEACH SCIENCE IN THE SECONDARY SCHOOL

Learning to Teach Science in the Secondary School is an indispensable guide to the process, practice, and reality of learning to teach science in a busy secondary school. Written by experienced teachers and expert academics, it explores core debates and topics in science education, providing practical and insightful advice with research and theory to support your development as a teacher.

This fully updated fifth edition focuses on the knowledge and skills you will need to develop your science teaching including key approaches to teaching physics, chemistry, and biology, lesson and curriculum planning, and assessment. There are also new chapters on:

- Safety in science teaching
- The science of learning for teaching science
- Mathematics and learning science
- Science for social justice
- Inclusive and adaptive science teaching
- Making use of research: practical guidance for science teachers

Written with university and school-based initial teacher education in mind and including learning objectives, lists of useful resources, and specially designed tasks in every chapter *Learning to Teach Science in the Secondary School* offers all student and early career teachers accessible and comprehensive guidance to support the journey of becoming an effective science teacher.

Lindsay Hetherington is Associate Professor of Science Education at the University of Exeter, UK.

Luke Graham is Lecturer in Education at the University of Exeter, UK.

Darren Moore is Senior Lecturer in Education at the University of Exeter, UK.

LEARNING TO TEACH SUBJECTS IN THE SECONDARY SCHOOL SERIES

Series Editors: Susan Capel and Marilyn Leask

Designed for all students learning to teach in secondary schools, including those on school-based initial teacher education programmes, the books in this series complement *Learning to Teach in the Secondary School* and its companion, *Starting to Teach in the Secondary School*. Each book in the series applies underpinning theory and evidence to address practical issues to support student teachers in school and in higher education institutions in learning how to teach a particular subject.

LEARNING TO TEACH SCIENCE IN THE SECONDARY SCHOOL

A Companion to School Experience

Fifth Edition

Edited by
Lindsay
Hetherington,
Luke Graham, and
Darren Moore

Routledge
Taylor & Francis Group

LONDON AND NEW YORK

Designed cover image: © Dani Pasteau

Fifth edition published 2024
by Routledge
4 Park Square, Milton Park, Abingdon, Oxon, OX14 4RN

and by Routledge
605 Third Avenue, New York, NY 10158

Routledge is an imprint of the Taylor & Francis Group, an informa business

First edition published by Routledge 1998
Fourth edition published by Routledge 2015

British Library Cataloguing-in-Publication Data
A catalogue record for this book is available from the British Library

Library of Congress Cataloging-in-Publication Data
Names: Hetherington, Lindsay, 1978– editor.
Title: Learning to teach science in the secondary school : a companion to
    school experience / edited by Lindsay Hetherington, Luke Graham, and
    Darren Moore.
Description: 5th edition. | Abingdon, Oxon ; New York, NY : Routledge, 2024. |
    Series: Learning to teach subjects in the secondary school series | Includes
    bibliographical references and index.
Identifiers: LCCN 2023056988 (print) | LCCN 2023056989 (ebook) |
    ISBN 9780367626686 (hbk) | ISBN 9780367626662 (pbk) |
    ISBN 9781003110187 (ebk)
Subjects: LCSH: Science—Study and teaching (Secondary)
Classification: LCC Q181 .L497 2024 (print) | LCC Q181 (ebook) |
    DDC 507.1/2—dc23/eng/20240222
LC record available at https://lccn.loc.gov/2023056988
LC ebook record available at https://lccn.loc.gov/2023056989

ISBN: 978-0-367-62668-6 (hbk)
ISBN: 978-0-367-62666-2 (pbk)
ISBN: 978-1-003-11018-7 (ebk)

DOI: 10.4324/9781003110187

Typeset in Interstate
by Apex CoVantage, LLC

CONTENTS

CONTENTS ■ ■ ■ ■

20 IS EDUCATION RESEARCH VALUABLE FOR TEACHERS OF SCIENCE?   265

LEE ELLIOT MAJOR

21 PUTTING RESEARCH INTO PRACTICE   277

JUDITH BENNETT, PETER FAIRHURST, AND ALISTAIR MOORE

FIGURES

TABLES

BOXES

TASKS

CONTRIBUTORS

Judith Bennett is Professor Emeritus at the University of York. Prior to this, she was the leader of the University of York Science Education Group (UYSEG) and Director of the Best Evidence Science Teaching (BEST) project. She was also head of the Department of Education at the University of York. Before joining the university, she spent several years as a high school teacher of science.

Richard Brock is a lecturer in science education at King's College London. After working as a secondary physics teacher, he now teaches on the science PGCE, MA, and doctoral programmes.

Ann Childs is Associate Professor in Science Education at Oxford University Department of Education. She is director of the Masters in Teacher Education, a masters course specifically designed to educate teacher educators and a science teacher educator on the Postgraduate Certificate in Education with a specialisation in chemistry. Her key research interests are in teacher education and teacher educator learning.

Paul Davies has been involved in education for more than 20 years. He is currently Assistant Head in charge of teaching and learning at Queen's College, London. Paul's role involves overseeing staff career professional development, the Early Career Teacher Framework and initial teacher education. He also teaches biology. Previously, Paul led the Science PGCE at UCL Institute of Education, where he remains an associate.

Lee Elliot Major OBE is Professor of Social Mobility at the University of Exeter. He was previously Chief Executive of the Sutton Trust. He commissioned and co-authored the original Sutton Trust-Education Endowment Foundation toolkit, a guide used by 100,000s of teachers across the world. He has a PhD in theoretical physics.

Peter Fairhurst is Physics Curriculum Specialist in University of York Science Education Group (UYSEG) and Co-Director of the Best Evidence Science Teaching (BEST) project. He taught in a variety of schools in England and overseas and was Head of Science at a high school in York before joining the University of York.

Caro Garrett taught sciences, mainly physics and chemistry, in secondary schools and a sixth form college for 25 years before joining the Education School at the University of Southampton in 2008. She introduced the pre-ITE 24-week physics Subject Knowledge Enhancement course and was the lead tutor for the Science PGCE and Programme Director for the Secondary PGCE programme. In semi-retirement, she worked part-time for the IoP as a School-based Physics Coach and for the local Science Learning Partnership.

Luke Graham was a science teacher for more than 20 years. He researches rural educational disadvantage with schools across the South West and is a member of the Centre for Social Mobility. Luke is the director for Secondary Teacher Education and the lead for Science initial teacher education at the University of Exeter.

Judith Hillier is Deputy PGCE Course Director and lead PGCE science tutor at the University of Oxford. She also teaches on the Masters in Learning and Teaching and the Masters in Teacher Education and is a Fellow of Kellogg College, Oxford. Judith's research interests lie in the education of science teachers, the recruitment and retention of physics teachers, the role of language in the development of scientific explanations in the classroom, and gender and diversity in STEM education. In 2021 she was awarded the Marie Curie-Sklodowska Medal by the Institute of Physics for her significant contribution to the support of women in physics.

Lindsay Hetherington is Associate Professor of Science Education and Deputy Head of Department at the University of Exeter, where she was also Head of Initial Teacher Education until August 2023. Lindsay conducts research into the use of materials alongside dialogue in learning science and into creativity in science education. Before joining Exeter, she was Head of Chemistry at a school near Bristol.

Andrew Howes leads Initial Teacher Education at the University of Manchester and is committed to collaborations in teaching, writing, and research which open up possibilities for relevant science education in the context of the climate and environmental crisis and social injustice.

Sam Mead is a qualified secondary science teacher who taught in Cornwall before moving to his current role as a software engineer. During his teacher education, he became very interested in evidence-based teaching and created the blog '52 papers', https://52papers.org/, with the aim of summarising a key research paper each week to inform practice.

Alistair Moore is the Biology Curriculum Specialist in University of York Science Education Group (UYSEG) and Co-Director of the Best Evidence Science Teaching (BEST) project. Prior to joining the University of York, he worked for one of the principal awarding organisations in England, where he still holds a chief examiner role for biology.

Darren Moore lectures in the School of Education at the University of Exeter. He has worked on the Secondary Science PGCE programme there since 2009. He also leads Masters modules on Special Educational Needs and Psychology. He has experience working with learners with SEN from primary age to further and higher education and been involved in a range of research involving learners with SEN, often focused on social and emotional mental health.

Jill Noakes has taught in secondary, further and higher education settings. She has delivered CPD and support for non-specialist teachers through the IOP's Stimulating Physics Network and was the subject tutor for PGCE Physics at the University of Exeter for several years while completing her PhD in education. Jill currently teaches physics at Queen Elizabeth's School in Crediton and is an examiner for Edexcel GCSE Physics.

Richard Osborne is a highly experienced educational technology consultant, researcher, and manager with over 30 years of experience across higher education and business. He has a strong track record of success in innovative educational technology solutions. He is also a qualified secondary school science teacher. He currently works as Faculty Learning Technology Lead for the Faculty of Mathematical and Physical Sciences at University College London.

Nick Pointer is Associate Dean of Learning Design at Ambition Institute, a non-profit teacher development organisation based in the UK. He is interested in how teachers, school leaders, and teacher educators can make sense of the latest research into learning, effective teaching, and teacher development, with a focus on supporting educators to explore how these ideas might translate into practice. Nick previously taught as Director of Science in state schools in London and the South East of the UK.

Michael J. Reiss is Professor of Science Education at University College, London; Honorary Visiting Professor at the Royal Veterinary College; Honorary Fellow of the British Science Associate; Fellow of the Academy of Social Sciences; a member of the Nuffield Council on Bioethics; and President of the Association for Science Education. After a PhD and post-doc in evolutionary biology, he trained as a secondary teacher and taught in schools before returning to higher education to teach on teacher education programmes and conduct educational research.

Stuart Ruffle has extensive experience in science secondary education in the UK that included school leadership roles. He has been Specialist Leader of Education with designations in Teaching and Learning, ITE, and CPD. Stuart has mentored trainees in school on six different programmes and has authored a Subject Knowledge Enhancement course. Prior to teaching he was a researcher and lecturer in the field of plant biological chemistry. He is currently Lecturer of Physics Education at the University of Exeter.

Ed Walsh is an author, curriculum developer, and CPD provider. A teacher for 20 years and subject lead for 12 of those, he then went into school improvement and ITT work. Ed is now an author and editor, writing for Collins and the ASE. He designs and presents CPD and is a Senior Facilitator for the National STEM Learning Centre and AQA Trainer.

Victoria Wong is a senior lecturer in STEM education at the University of Exeter where she is the subject tutor for PGCE chemistry. She previously worked as a teacher educator at King's College London and the University of Oxford and taught chemistry and science for 12 years in England, Spain, and New Zealand. Her research interests include students' use of mathematics within science and how science education policy is made and enacted.

ACKNOWLEDGEMENTS

We would like to thank those authors who have contributed to previous editions of *Learning to Teach Science in the Secondary School* whose work has provided the foundation for this edition. Rob Toplis, Katherine Little, Steven Chapman, Ralph Levinson, Kevin Smith, Pete Sorensen, Marcus Grace, Ruth Amos, Sandra Campbell, Christine Harrison, and John Oversby.

We would also like to thank Susan Capel for providing us with the opportunity to edit this new edition and for her support throughout the process.

INTRODUCTION

Lindsay Hetherington, Darren Moore, and Luke Graham

Science is about being **curious** about the world and wanting to **know** more about how it works. It is about fascination, innovation, creativity, and knowledge. Often science teachers are attracted to the career precisely to share this wonder about the world (both in terms of awe and curiosity) with young people. The skills that make them good scientists are also important in their becoming good teachers: Curiosity about what makes children tick, what their interests are, and how to engage them in learning; knowledge about both science and how to teach science effectively; always learning and finding out more to become better teachers; being keen to innovate and find good ways to teach core concepts. All of these are things you will need as you move through your initial teacher education (ITE) and on into a (hopefully) long career in teaching.

Teaching is a fascinating and complex activity that involves working with some of the most interesting human beings on the planet. As a science teacher you will need essential teaching skills, excellent knowledge of your subject, and the capacity to be responsive and adapt to the needs of the pupils you are working with. You will need to use your professional judgement to make decisions about the best pedagogical approach to use to teach *these* objectives to *these* pupils. Having a good understanding of key theories in education and the research that underpins them is helpful to you in enabling you to exercise that judgement. Critical appraisal of educational research methods and approaches will also help you stay up to date as you move through your career. Your initial teacher education course should introduce you to all the elements you will need, providing a foundation for your career as a teacher. It is likely that your course will scaffold this learning for you through a sequenced ITE curriculum that enables you to develop the basic knowledge and essential skills you need to start with. It will help you to be able to reflect on and evaluate your teaching with the support of an expert teacher such as your mentor. You will then gradually develop more complex skills rooted in knowledge about broader elements of teaching, such as

DOI: 10.4324/9781003110187-1

adaptive teaching for a range of pupil needs. Throughout, it is helpful to try to see the practical-skills and knowledge-based elements of your course as woven elements of the same tapestry, rather than as separate 'chunks'. Knowing more about the fundamentals of science education and how that knowledge has been developed through research and practice will help you develop your own practice. Likewise, developing your practice will help you understand and interpret educational knowledge and understanding of theory and research. Your ITE course will organise your school experience placements and your centre-based learning in different sequences. They may scaffold your experience differently, for example, beginning with teaching in controlled, small-group or peer environments or setting you off with your own classes straight away. Regardless of this sequence, looking for ways to connect your developing knowledge and skills is key to enabling you to develop as an effective science teacher. In this book, we provide you with background information about teaching and science education which you can use alongside your course materials and school experiences to facilitate this connection of knowledge, theory, research, and practical skills.

ABOUT THIS BOOK

This book contains 21 chapters which cover important information about teaching and learning in science. The book begins with key information about general principles about what to expect from your initial teacher education and some approaches that will support you settling in. We then move on to focus on core aspects of the science curriculum, with chapters about keeping your pupils safe, pedagogical approaches to teaching the three main science subjects, and how to think about planning to enable your pupils to progress in their learning. Following this is a set of chapters that cover key topics in science education that are based in research evidence about what makes good science teaching. These chapters aim to introduce important knowledge from research whilst enabling you to make links to your practice. They include ideas about how cognitive science can be used to inform pedagogical practices, how pupils construct knowledge and the implications of this for teaching, and the role of models and modelling in science education. We explore how to make the most of learning technologies and practical work in science education. Then we discuss how language and literacy and mathematics connect with science education and how teachers can support pupils with these skills. Two chapters focus on the role assessment plays in science education. Assessment is crucial as it enables teachers to adapt teaching in response to pupils' needs, which is covered next alongside key issues for inclusive science teachers. We conclude this set of chapters with one that discusses the role science education can play more broadly, for example in relation to social and climate justice. We end the book with two chapters that highlight the key role of research in science education, which is foundational to your continued development as a teacher across your career. As you will see from this brief outline, the book does not aim to provide you with the subject and curriculum knowledge you will need to teach science effectively. For this, you will need to draw on your own subject knowledge from your earlier studies and also refer to other sources of information to develop this knowledge, such as the useful resources available from examination boards, the Association for Science Education, Royal Society of Biology, Royal Society of Chemistry, Institute of Physics, and the Earth Science Education Unit.

In this book, each chapter is laid out as follows:

■ *Introduction* to the content of the chapter
■ *Objectives*, presented as what you should know, understand, or be able to do when you have read the chapter and carried out the tasks
■ The *content*, based on research and evidence to emphasise that teaching is best developed by being based on evidence and critical reflection. The content is interwoven with *tasks* to aid you in connecting the ideas in the chapter to your own experience and practice
■ *Summary and key points* of the chapter
■ *Further resources*, selected to enable you to find out more about the content of each chapter.

The kinds of activities encouraged through the tasks in each chapter are designed to enable you to undertake inquiries into teaching and learning science in your context. They might ask you to reflect on reading, observe teaching and learning, or discuss key ideas with other student teachers or with a mentor or tutor. Some of these tasks involve you in activities where you will be asking something of another busy teacher, technician or teaching assistant, member of senior leadership, or another busy student teacher. For example, it might suggest you observe a teacher's lesson for a particular purpose. If a task requires you to do anything that impinges on another person, *you must ask permission of the person concerned*. You should always behave with the utmost professionalism and take an ethical stance in your approach to tasks, including considering issues of confidentiality and sensitivity in how you use personal information that is shared with you. It is crucial that you respect the school in which you are working and the people with whom you work. The tasks we have included are designed to challenge you to be critically reflective about aspects of teaching to help you develop your own ideas about teaching, but this is not the same as criticising others, so be thoughtful and considerate of others throughout. You may be able to use the tasks as evidence of progress towards any teaching standards you may be working towards in the jurisdiction in which you are based. Similarly, the tasks may support you in developing your work at the masters level (M level). Many teacher education courses include accreditation at M-level and earn you some credits towards a master's degree which can be taken forward into further study to enable you to gain a full master's.

The text is also supported by other books in the series, including:

■ Newton, D. P. (2023) *A practical guide to teaching science in the secondary school,* Abingdon: Routledge.

There are also two generic texts:

■ Capel, S., Leask, M., Younie, S., Hidson, E., & Lawrence, J. (Eds.). (2023) *Learning to Teach in the Secondary School: A Companion to School Experience*, 9th Edition. Abingdon: Routledge. You can also access material on the companion website for the generic *Learning to Teach text (www.learningtoteach.routledge.com)*
■ Capel, S., Leask, M., Lawrence, M., & Younie, S. (Eds.). (2020). *Surviving and Thriving in the Secondary School: The NQTs Essential Companion*. Abingdon: Routledge

Finally, though primarily written for mentors, some of the texts and tasks in the following book would also be useful for student teachers:

■ Salehjee, S. (Ed.). (2021) *Mentoring Science Teachers in the Secondary School: A Practical Guide*

A NOTE ON TERMINOLOGY

In this book, we are referring to ITE generally, and it should be relevant and useful to student and early career science teachers regardless of where they are undertaking their ITE. However, although ITE is generally referred to generically, where we do refer to specific requirements and curriculum examples, we mostly refer to requirements and policy in England. If you are not learning to teach in England, you should refer to the specific requirements of your own ITE or your own national curriculum or policies at this point. We recognise that you may not be undertaking ITE that prepared you to teach in state schools in England, and we suggest that for the information and tasks given, you substitute relevant curriculum and requirements from your own situation and reflect on the differences between the curriculum and requirements mentioned in the chapter or task and those in your specific context and situation.

Your ITE provider may use documentation to support you in your development and enable them to assess your progress with respect to their ITE curriculum. They will have specific terminology they use to refer to such documents or portfolios such as teaching file, professional development portfolio (PDP), individual development portfolio, reflective journal, etc. The tasks in this book may be helpful for you to record in your PDP (or similar) and may be useful to reflect on as you identify strengths and areas for development throughout your ITE and in your transition to your early career.

In this book, we use the term *pupils* to refer to school children, and the term *students* to refer to those learning in further and higher education. We use the term *student teachers* for people who are undertaking ITE. We refer to *ITE* as we would argue that your ITE provides not just training (ITT) but also education (see Chapter 1 for more on this point). At root, this is because education is about more than just being trained in a specific set of skills but is about becoming somebody new: a teaching professional.

You will be supported through your ITE by key staff both on your school placement and from your ITE provider (the central team that design and quality-assure your ITE curriculum and coordinate and assess your learning, which may be a university, school-led provider, or other company registered as a provider of ITE in your context). We will refer to the *mentor* as the key person who supports you during school experience, usually an experienced teacher in your department. We will use the term *tutor* to refer to your key member of ITE provider staff who supports you. You may have a *subject tutor* and a *pastoral tutor* linked to your provider: we use the general term *tutor*.

Note that the terminology in this book may be different to that used in your ITE. You should check terminology used in your ITE and adjust accordingly for your needs when completing tasks you might use in your own individual portfolios and assignments.

FINALLY

We hope that you find this book a really helpful introduction to key ideas in science education. Our aim is to support you in your development as a teacher and support you in maximising your pupils' learning. We welcome feedback about what is helpful or not, so please do let us know.

We wish you a fulfilling, enjoyable and productive career as a science teacher and hope you find it full of fascination, curiosity, innovation, knowledge . . . and fun. Good luck.

Lindsay Hetherington
Luke Graham
Darren Moore

BECOMING A SCIENCE TEACHER

Caro Garrett and Lindsay Hetherington

INTRODUCTION

The role of a science teacher is demanding, encompassing a wide range of skills and situations: it is often conceptually difficult and pupils may come to science lessons with a variety of preconceptions and previous learning experiences. Science teachers must deliver safe and engaging lessons involving relevant practical work. Additionally, they are expected to cover the science curriculum at an appropriate level, provide concise and accurate explanations of often complex concepts, assess pupil learning and progress over time, and all the while, inspire and motivate our future scientists. As you read this you will probably already have started on your route to becoming a qualified secondary science teacher and may be in the early stages of your career. You will be increasingly aware of the range of skills, knowledge, and understanding needed to be a successful science teacher. Even with a set timetable, no day – or hour in the day – is the same. You are dealing, on a daily basis, with a unique set of young people, in an age group that is going through some of the most important changes of their lives. As a secondary science teacher you are in a privileged position to witness and, to some extent, to be part of those changes. Therefore, the skills of a student teacher or early career science teacher are not those that just involve scientific knowledge and skills but also skills of empathy, effective communication, and performance. You may very well be joining the teaching profession because you understand the expectation that you can develop this great range of skills and relish the opportunity to do so.

The philosopher of education Gert Biesta wrote about three parameters around which the purposes of education could be framed, namely, qualification, socialisation, and subjectification (Biesta, 2015). Qualification is about ensuring those educated have the necessary knowledge, skills, and understanding to do something. An example would be the knowledge, skills, and understanding you

DOI: 10.4324/9781003110187-2

will develop in order to teach science, enabling you to meet the Teacher's Standards (DfE, 2011) by the end of your course. Socialisation is about becoming members of a particular social or cultural group, understanding the norms and values of that group or institution. This is not always explicitly taught but it is an important part of education. In the context of becoming a science teacher, for example, socialisation is about understanding the norms of how schools function, such as how to work with a science technician. Finally, subjectification is about how education contributes to an individual's development *as an individual* – how it changes who they are and enables them to become autonomous and independent in their thinking and acting. The sections in this chapter are organised around these three functions of your initial teacher education. First, they provide some background relating to starting points and suggest how you might audit the skills and knowledge you may bring with you and orient you towards what you will need to learn to qualify as a science teacher. Second, it focuses on information about how schools and science departments are structured and where you as a science teacher 'fit in'. Third, it discusses how you can support yourself as you become an independent, autonomous science teacher.

OBJECTIVES

At the end of this chapter, you should be able to:

- Understand the kinds of knowledge and skills you will need to develop as a science teacher
- Identify your starting points
- Explain how science departments in secondary schools are typically organised
- Form appropriate and constructive professional relationships with others in the science team and wider school
- Begin to view your first years in teaching as part of a progressive model of development for a long-term career in teaching.

KNOWLEDGE, SKILLS, AND UNDERSTANDING

Starting points: What do you know already?

Student teachers come from a wide variety of starting points in terms of their academic experience, social and cultural experiences, and work experiences. Added to this are their values, attitudes, and beliefs about science, what it is, and how it should be taught.

Academic experiences may be varied. They may include a first degree from a fairly narrow area or one with a mixture of different modules; they may include a higher degree in an even narrower area with research based on one specialist topic. Examples may be a biology student teacher with a first degree in neuroscience but with no ecology; a physics student teacher with a degree in electrical engineering but with little content in astrophysics; or a chemistry student with a degree in forensic science but little physical chemistry. In these examples, further subject knowledge enhancement would be required before being able to confidently teach all aspects of the specialist science. There are other degrees which may be considered for acceptance into science teacher education courses, such as geology, sport science, archaeology, etc., which may require an extended subject knowledge enhancement course.

An Individual's social, cultural, and work experiences can often be a valuable addition to the daily interactions with pupils. Personal experiences and interests, memberships of groups, travel experiences, and hobbies can contribute to the positive professional relationships that occur between teachers and pupils. At one level, involvement in the clubs and societies in schools not only helps forge these positive interactions but helps the informal education of pupils: the hidden curriculum. At another level, the richness of a diversity of backgrounds and cultures can add to the overall pupil experience in school. Work experiences may contribute to new ideas to teaching science, approaches to organising the classroom, dealing with individuals – or simply some of the anecdotes from work that can be used to illustrate ideas in the science laboratory.

Task 1.1 **Starting out: your characteristics**

As you start out on your teaching journey, it is useful to make a list of your characteristics (knowledge, skills and beliefs) related to science teaching and learning.

These should include: subject knowledge, curriculum knowledge, knowledge about how schools work, knowledge about school examinations, observational skills, communication skills, mathematical skills, and modelling skills.

These could include: 'transferable' skills such as organisation, time management, and creativity; 'people' skills such as empathy, diplomacy, enthusiasm, and beliefs; and attitudes and values that might address the question, 'why do I want to teach science?'

Make your list and then reflect on your areas of strength and areas to develop. Think about how you can enhance these skills and how you hope to address some of these areas during your teacher training and education.

Subject knowledge, content knowledge, and pedagogy

There has been a certain amount of debate about the nature of subject knowledge. Teachers need to know *what* to teach, the content knowledge necessary. They also need to know *how* to teach this knowledge, the pedagogy involved. Shulman (1986) has contributed to our understanding about subject knowledge and has proposed the term *pedagogical content knowledge*, or PCK, to refer to the practical knowledge used by teachers in classrooms. This practical knowledge is, understandably, complex as it involves the knowledge that specialist teachers possess that includes pupil misconceptions, examples, analogies, and models. Added to this are the illustrations, conceptual difficulties, and connections with other aspects of learning such as assessment and the curriculum (Berry, 2012). If we take the example of teaching a very simple topic such as the forces on a cyclist pedalling at a constant speed along a flat road, the teacher will need to know a number of important facts. They will need to know the content knowledge about the forces acting on the cyclist such as friction, air resistance, and gravity and about Newton's Laws. They will also need to know pupils' potential misconceptions or alternative frameworks about forces and motion, how force arrows can be drawn and used to represent balanced forces, some possible simple demonstrations or observations about Newton's Laws, other possible examples that can add to pupils' understanding, 'what if' questions, and even the kinds of questions that may arise in assessment tests or examinations. The PCK involved in this apparently straightforward example on forces and motion is rather more complex than it immediately appears, and the teacher needs to draw on a wide range of knowledge to deal with this.

Task 1.2 **Pedagogical content knowledge (PCK) for teaching simple photosynthesis**

To help you understand the wide range of things to consider when planning a lesson, list all the items of PCK you can think of that would be needed to teach a simple outline of photosynthesis, involving the production of carbohydrate and oxygen from carbon dioxide and water, using light energy. Use the considerations given earlier in the example of forces on a cyclist, to help you determine how you could represent and formulate the subject content to make it understandable for your pupils. Apart from subject knowledge, this would include potential misconceptions, simple demonstrations, and modelling you could use to illustrate concepts and facts and questions you could ask. Start to embed this approach to preparing your lessons in the future.

Curriculum knowledge

Teachers need to have detailed subject knowledge, but they also need to know *what* needs to be taught and *when* i.e. curriculum knowledge. This is further complicated by the frequency of curriculum change, but change is inevitable as the curriculum is revised in response to changes in policy and evolving ideas about what kind of science needs to be taught to all pupils in the secondary age range. Curriculum change occurs throughout the world as governments and international educators react to the need for both a scientific and technological workforce while at the same time enhancing the scientific literacy of twenty-first-century populations who need to be better informed about some of the major scientific, ethical, and environmental issues facing them.

The National Curriculum for England (initially including Wales) arrived for first teaching in schools in 1989, resulting from a mixture of historical events, initiatives, and a not inconsiderable degree of political influence. Since 1989 there have been five further versions of the National Curriculum, first taught in 1995, 2000, 2008, and 2014 (DfE, 2015). Whenever the National Curriculum changes, teachers have had to respond by producing new schemes of work to accommodate all the factors, and they can expect to be inspected to ensure that they are delivering this curriculum.

Many countries have some sort of school inspection system in place. Ofsted is the non-ministerial department in England that is responsible for inspecting services providing education and skills for pupils of all ages. The other devolved nations have similar inspectorates in place (Education Scotland, Estyn (Wales) and The Education and Training Inspectorate (Northern Ireland)), all with the same fundamental purpose: to improve the overall quality of education and training for all learners. As an example of their vision, in the most recent Education Inspection Framework published by Ofsted (2019), it is clear that the quality of education will be judged through a focus on curriculum and the aspirations schools have for their learners. The buzz words 'Intent', 'Implementation', and 'Impact' are used to emphasise that the inspectors will be looking at how broad, coherent, and challenging the curriculum is, with knowledge, skills, and cultural capital all playing a part in this.

Beyond the UK, in recent years, there have been continuing international concerns about school science education, including a reduction in the numbers of pupils studying the physical sciences beyond the age of 16, gender differences, and pupils' attitudes and motivation for studying science. The Relevance of Science Education (ROSE) study of pupils' attitudes to science shows that 'in over 20 countries, pupils' response to the statement: 'I like school science better than other subjects' is increasingly negative the more developed the country' (Osborne & Dillon, 2008, p. 13) and that science is 'important but not for me' (Jenkins & Nelson, 2005, p. 41). The ASPIRES project is a large longitudinal UK study of young people's science and career aspirations, which is seeking 'to generate new understandings of how and why young people come to see science as being 'for me', or not, with the goal of informing policy and practice to support increased and more equitable participation in STEM' (Archer et al., 2020, p. 4). Against this backdrop has been the most recent version of the English National Curriculum with greater emphasis on content knowledge. It remains to be seen if this initiative is able to reverse some of the trends in attitudes to school science and can engage *all* pupils in further study and for greater scientific literacy.

WORKING AS PART OF A SCIENCE TEAM

The majority of your experience will be spent in schools and working in a science department. This will involve daily interactions with science teachers, science technicians, form tutors, teaching assistants, and cover supervisors as well as occasional interactions with other school staff such as caretakers, maintenance staff, visitors, and the school's senior leadership. In addition, you will also be working within pastoral or year group teams that will include teachers from a range of subjects and pastoral team leaders and possibly the Special Needs Coordinator (SENCo): teaching in schools is not just about teaching your own subject. This section will give you an overview of school organisation and will help you explore the ways in which a science team works in your placement school - in other words, it will help you understand what you will be learning to become socialised into schools as a science teacher.

Moving from the primary school to the secondary school

You are likely to spend some time either before or during your course in a primary school (age 4-11) and therefore will be in a position to appreciate the very different ways in which the schools operate in these two phases of education, both on a structural and on a daily basis. If you do not have the opportunity to do this as part of your course, consider arranging independently to visit a primary school to see how learning is organised and what pupils experience. When pupils move from primary to secondary schools they frequently experience enormous changes. It is important for you to consider this when teaching your year 7 classes.

> Moving from primary school to secondary school is a very exciting but sometimes daunting experience for many students. When the transition from Year 6 to Year 7 is right, we see students get it right; we see students in Year 7 who are engaged, interested and enthusiastic about science; students who are keen to learn and at the same time continue to make progress in the subject.
>
> (Warren, 2019, p. 9)

While observing in a primary school you are likely to be given specific tasks to focus your observations on the differences in teaching strategies, learning activities, pupil motivation, school structures, the classroom environment, etc. This can be an extremely enjoyable and enlightening experience, so make the best use of it! Collins and Reiss (2016) have edited a themed edition of the Association for Science Education's (ASE) School Science Review which contains a number of informative articles about the primary to secondary transition.

Organisation in schools and the science department

Many, if not all, schools in the UK and in other countries follow a hierarchical model with a single Headteacher at the top of the pyramid, followed by Deputy Headteachers and Assistant Headteachers (the Senior Leadership Team) and below this the teaching teams based around subjects and year groups (with pastoral responsibilities). Schools that are part of multi-academy trusts (MATs) will have an executive layer of leadership above this, managing the group of schools working together. The typical structure of schools and departments reflects the need to respond to external pressures, to get a multitude of tasks associated with teaching and school organisation completed, a need to provide incentives to do so, and at the same time allowing teachers to see a route to personal and career advancement.

The science department may be organised in a number of different ways, depending on whether the hierarchy is based on the separate sciences, by age group (Key Stage in England), or as part of a wider Faculty. UK schools with post-16 provision, where subject specialists are needed, may have Heads of each science, but those without post-16 provision may be more likely to have a Head of Science and teachers within the department with individual responsibilities such as Heads of Key Stage KS3 (ages 11-14) and KS4 (ages 14-16). Within all these structures there may be graded posts of responsibility, Teaching and Learning Responsibility points, or Lead Practitioner points, with associated payments. These points could be for responsibilities for the KS3 curriculum, the KS4 curriculum, assessment and recording within the department, being in charge (i/c) of a separate science at KS4 and A level, coordination of required practicals, and so on. There may be other roles taken on by science staff, outside the department, such as whole school responsibilities for work experience, careers or examinations, co-ordinator for continuing professional development (CPD) and, relevant to student teachers, a role of responsibility for student teachers and early career teachers (ECTs) in the school.

Task 1.3 **Identifying who's who in the school**

You are now joining the school staff and one of the early tasks you will need to do is to find out how staff teams are organised. This information may be in a staff handbook so it is worth finding this out.

1   Who is the Headteacher? Who are the Deputy Heads and Assistant Heads and what are their main responsibilities?
2   Who has responsibility for leading Initial Teacher Education (ITE) in the school?
3   How are subject or curriculum teams organised?
4   How is the pastoral system organised?
5   Which pastoral team will you be part of? Does this include members of the science team?
6   What school meetings and committees are there?
7   Who is the Special Educational Needs Co-ordinator (SENCo)?

Support staff in the science department

Technicians are often (sadly) overlooked outside the science department, but they are some of the most important people in your science teaching life. Technicians are rarely noticed as they move through and between laboratories, but they see everything. During your first day or two, it is very important to talk to the technicians and find out how the apparatus, the provision of books and worksheets, stationery, and photocopying are organised. Are there class sets of books and do you have to order them? Is there a number for the photocopier? Are there teaching resources on the school computer system, and how do you access them? How do you get class sets of copies reproduced?

Task 1.4 **Finding out about the role of your technician**

One way of finding out exactly what sorts of jobs a technician does is to observe them and talk to them. This can be recorded in a notebook or diary and referred to a later date. Your records may form part of your portfolio for fulfilling your wider professional responsibilities. Some questions may be:

1 When should you wear a lab coat and safety glasses?
2 When should pupils wear a lab coat (if at all) and safety glasses?
3 When should you or your pupils wear goggles rather than safety glasses?
4 What should you do in the event of a fire?
5 Where are the emergency gas and electricity shut-offs?
6 How is the record kept of laboratory equipment?
7 How is the record kept of chemicals?
8 How is the apparatus ordered for each lesson, and how much notice is needed?
9 How are the equipment and chemical stores organised?
10 Where does the technician place apparatus for lessons?
11 What are staff expected to do with equipment at the end of lessons?
12 How is broken glass from the classroom disposed of safely?

The procedure for ordering apparatus and chemicals is one of the most important things to find out early. In particular the following points are important to clarify with the technicians:

Technician order forms. Does the science department use a standard order form? If so, it may be a good idea to keep an example for your portfolio.
How does the risk assessment work? It may be that the scheme of work used by the department includes specific risk assessments for particular practical work activities. However, you will also need to make sure that you can demonstrate that you can assess risk, are aware of the possible hazards

with different age groups and classes, and can show this in your evidence (see Chapter 3 for more on safety). An important reference for support with this is the CLEAPSS website.

How much notice do the technicians require? Frequently this is a week, but different technicians in different schools may have other arrangements. It will be up to you to find out about this and keep to the notice periods. You cannot rely on coming into a prep room on a Monday morning with a list of apparatus needed for later that day or, worse still, the first teaching period.

You will need to specify exactly what you want for apparatus and materials; you cannot assume they will be there in the laboratories. It is also inappropriate to just order 'beakers' for a practical – you will need to specify size and number. Similarly just stating 'hydrochloric acid' is not enough. How much? In how many bottles? At what concentration? Ordering materials also applies to nonlaboratory items as well. It is very easy to forget basic things such as a class set of pencils, writing paper, glue sticks, or scissors.

The technician can be one of the most important allies of a student teacher, so it is important to treat them with due respect and gratitude!

The second group of people you will need to consider in your planning, teaching, and learning is the Teaching Assistants (TAs), also called Learning Support Assistants (LSAs) or Classroom Assistants (CAs). Different schools may have different numbers of TAs and they will have different roles. In a number of cases they will be employed to support specific pupils, in other cases they may have more of a whole class role. TAs are colleagues with whom you work and it is important that they are included in your planning and teaching and that you develop some means of informing them of your plans prior to the lessons in which they will be supporting pupils. Your work with TAs can be regarded as part of the Teachers' Standards (DfE, 2011) where you are required to develop effective professional relationships with colleagues, knowing how and when to draw on advice and specialist support and how to deploy support staff effectively. It is therefore important that you are not only able to evidence this but also to take a proactive approach to involving TAs in your work on a daily basis. For more on working effectively with teaching assistants, see Chapter 19.

BEING A STUDENT TEACHER IN A PROFESSIONAL SETTING

Working at master's level

In England and Wales, as part of your development as a professional teacher of science, there is usually an expectation that you are able to work at an academic level that meets master's level criteria. Many initial teacher education courses, whether they are university- or school-based, have assignments credited at master's level (M-level) which lead to the award of a PGCE. This qualification is distinct from – but often woven together with – the Qualified Teacher Status (QTS) gained from meeting the Teacher's Standards. M-level assessment criteria outline the level of thought and writing needed for work at this level which encompasses theory, critical analysis, reflection, and enquiry. As part of this M-level work you will very likely be asked to focus on an aspect of science teaching from your own school experience and may be asked to prepare and carry out a small-scale project in school that collects and analyses qualitative or quantitative data and is informed by some theoretical perspective. The utility of educational research for teachers is discussed in more detail in Chapters 20 and 21. Drawing on M-level critical thinking and analytical skills is part of a long-term teaching career, enabling you to continue to develop yourself as an evidence-informed teacher in response to changes to policy, curriculum, society, and other circumstances.

Moving into your early career

As you complete your initial teacher education and move into your early career, you are likely to continue to be supported in various ways. You will typically have a slightly reduced teaching timetable and ongoing mentoring by an experienced teacher to help you continue to hone your knowledge and skills as you take up your first teaching post. In England, the government introduced the Early Career Framework (ECF) as a result of the need (accepted by all stakeholders) for a longer and more defined induction process for Early Career Teachers (ECTs).

The ECF underpins what all early career teachers should be entitled to learn about and learn how to do based on expert guidance and the best available research evidence. The ECF has been designed to support early career teacher development in five core areas – behaviour management, pedagogy, curriculum, assessment, and professional behaviours. (Department for Education, 2019, p. 5)

The ECF was designed to link directly to the Core Content Framework (CCF) that describes what is seen as essential knowledge and skills that should be taught in an ITE course. As part of your development across your ITE and early career, you will be learning about yourself and who you are as a teacher, as well as learning a wide range of knowledge and skills. You will need to develop a certain resilience, as teaching can be a demanding (albeit rewarding) job. Resilience is supported by both individual and social or contextual factors (Mansfield et al., 2016) and as you move through your ITE and early career, you will begin to reflect on and understand yourself and the individual, social, and school contexts in which you best thrive as a science teacher. As you are a unique individual, you will not teach in exactly the same way as someone else and indeed, since teaching is about communication and relationships as well as knowledge, you will need to draw on your own personality and experiences to teach well. This means that as you become a science teacher, you will change and learn, as well as teach. This relates to Biesta's final function of education – subjectification. Ongoing skills of reflection and dialogue with experienced teachers, mentors, tutors, and your peers will go a long way to helping you develop and thrive in your career as you become a science teacher, and this is discussed in more depth in Chapter 2.

SUMMARY AND KEY POINTS

This chapter has introduced some of the considerations that need to be taken into account when learning to be a science teacher. It has provided an overview of the complexities of science teaching and learning and of the professional role of science teachers. We have explored the range of different ways that schools are organised, the roles, responsibilities and diversity of different individuals in the science department and ways in which the student teacher can appreciate and work effectively in this environment.

Some of the areas covered are:

▪ The learning and skills you bring to science education as starting points
▪ Subject knowledge, pedagogy, and pedagogical content knowledge (PCK)
▪ The nature of curriculum knowledge
▪ The changes that pupils experience as they move from primary to secondary school
▪ The nature of being a professional and the professional relationships that science teachers forge.

Your progress across your initial teacher education will be marked by a shift from a focus on your own skills and understanding towards a time when your focus is entirely on the pupils' learning. Becoming a teacher takes stamina and persistence; it requires good administrative skills as well as skills and knowledge needed for the classroom. By the end of your initial teacher education you will be aware that you have learned 'a new way of being yourself' (Black, 1987) and will derive a great deal of professional satisfaction from this 'new you'.

Check which requirements for your initial teacher education you have addressed through this chapter.

FURTHER RESOURCES

Boxer, A. (2021). *Teaching secondary science: A complete guide*. John Catt.

This is an excellent book to support your development of pedagogical content knowledge. It offers clear outlines of key concepts and how you might teach them.

Capel, S., Leask, M., Younie, S., & Lawrence, J. (Eds.). (2022). *Learning to teach in the secondary school: A companion to school experience* (9th ed.). Routledge.

This general book for new teachers includes chapters on Becoming a Teacher, Beginning to Teach, The School, Curriculum and Society and Your Professional Development

CLEAPSS Science website. Retrieved March 29, 2024 from http://science.cleapss.org.uk/

CLEAPSS resources give teachers ideas for exciting and engaging practical activities that fire pupils' imaginations and then, unlike many other sources of ideas, go on to show teachers and technicians in detail how to translate the ideas into safe experiences in the classroom.

Driver, R., Rushworth, P., Squires, A., & WoodRobinson, V. (2004). *Making sense of secondary science*. Routledge.

This important book provides a research digest of many of the conceptual challenges in science education. It is an important part of background reading, as well as preparation and reference for lesson planning.

Scaife, J. (2018). Learning in science. In J. Wellington & G. Ireson (Eds.), *Science learning, science teaching* (4th ed.). Routledge.

Jon Scaife's extended chapter in this book provides an excellent introduction to learning theories with practical examples.

USING REFLECTION, EVALUATION, AND METACOGNITION IN LEARNING TO TEACH SCIENCE

Caro Garrett

INTRODUCTION

Reflective practice is a crucial aspect of learning to teach. You will engage in a 'plan-do-review-learn-apply' cycle each time you teach, reviewing, reflecting on, and evaluating your teaching and applying what you have learned to your next lesson. Your progress towards becoming a 'reflective practitioner' (a term introduced by Schön in 1983) is of fundamental importance to your development as an effective teacher not only during your initial teacher education (ITE) programme but well beyond. In fact, it could be argued that it is key to a satisfying and successful long-term career as a teacher. Honest and critical self-reflection following any lesson can be of great benefit both to your teaching and your pupils' learning, but this has to be a balanced and constructive activity: self-reflection is not about 'beating yourself up' after every lesson but about careful consideration of how you can improve your teaching on a day to day basis. This chapter will look at the various ways in which reflection and evaluation can be carried out with a focus on positive outcomes and encouraging progress in your practice.

OBJECTIVES OF THIS CHAPTER

At the end of this chapter, you should be able to:

■ Understand the role of reflective conversations with mentors, tutors and colleagues
■ Recognise ways in which you can help to participate in and enrich these conversations
■ Start to develop some effective personal reflective practices.

DOI: 10.4324/9781003110187-3

REFLECTIVE PRACTICE

The terms 'reflective practice' and 'the reflective practitioner' were first coined by Dewey (1910) and Schön (1983). Dewey wrote that being reflective 'enables us to direct our actions with foresight. . . . It enables us to know what we are about when we act'. (Dewey, 1910, p. 6). Reflection involves looking at your own actions, choices, and behaviours, their effects on your pupils' learning experience, and using these reflections to develop and improve your practice. While these statements are key to successful teaching, your ability to do this at the start of your practical teaching will be limited by your lack of experience, and it is important that you develop your competence in this aspect over time. Schön (1983) introduced the term 'reflective practitioner' and presented two aspects: 'reflection-in-action' and 'reflection-on-action'. Reflection-in-action refers to reflecting on and reacting to the situation in front of you in the classroom, responding to your observations of your pupils and changing what you are doing. For instance, you might choose an alternative model to demonstrate a difficult science concept or present a different explanation to help your pupils' understanding. This is an advanced skill which experienced teachers are able to use very constructively, but it requires you to have developed a variety of strategies and an understanding of different ways of presenting specific topics. Reflection-on-action (possibly referred to as reflection-for-action) occurs when you consider your lesson after the event and have the opportunity to reflect on what happened in your classroom. For instance, you might think about why your pupils did not understand something you were teaching and the alternative ways in which you could have delivered the topic.

Effective reflective practice can be a powerful learning tool, helping you to synthesise, analyse, and explain your experiences, whether from observing other teachers or considering your own teaching. From your reflections you may be able to make more sense of what happened in the classroom and use this understanding to develop your planning, strategies, explanations, and classroom management.

While we might reflect briefly on our actions in the normal flow of our lives, for reflection on your teaching to be effective, it needs to be in-depth and can initially feel weird and unusual. Teachers who operate in a culture of reflective teaching were better able to cope with the new approaches to teaching required during the COVID-19 pandemic when much teaching switched to online delivery. Effective reflection usually means writing things down and thinking about them and talking to people about it – critical reflection takes time and is an odd process the first time you do it, but is effective in supporting and accelerating student teacher development in ITE. The following sections introduce the role of reflection and evaluation in a planning and teaching cycle and then suggest some ways in which you can develop your ability to reflect constructively.

Task 2.1 **Reflections on your own experiences of learning science**

As a start to your journey as a reflective practitioner, think back to your own school-ing and recall an educational experience, preferably in science, which you remem-ber and value. Why was this a successful learning experience for you? List all the points you remember as being distinctive about this experience. With other science student teachers, discuss and compare your responses to this task. Collate the positive features identified into a list of the qualities a good teacher of science has.

Next, repeat the exercise, this time focussing on an educational experience that failed to engage or motivate you. Once again, compare and collate your responses with those of fellow student teachers.

Keep your notes in your Professional Development Portfolio (or similar) and at the end of the course, consider how much your ideas have developed and how much you have achieved.

PLANNING, MONITORING, AND EVALUATING YOUR TEACHING

Plan-do-review-learn-apply

A useful model which can be applied to the process of critically reflecting on practice and using that knowledge to improve teaching is that of an active learn-ing cycle represented in Fig 2.1 below:

This cycle involves:

- Planning: deciding what to do (possibly as a result of an iteration of the cycle)
- Doing: the carrying out of the plan, which, as well as delivering a lesson, could include other ways of getting ideas and information to improve prac-tice such as observing other teachers' lessons, including in subjects other than science, working with a tutor group, marking pupils' work, etc.
- Reviewing: standing back, describing the event, analysing it and evaluating its effect. The ability to reflect constructively and self-critically is essential. Such reviewing can be highly effective if done with another person, often a teacher in school or your tutor or a fellow student teacher (see the next section on reflective conversations). In the review, other ideas are brought in to make sense of what has happened and to help in identifying general points that can be applied elsewhere. It is in this phase that links are made to the more theoretical parts of the course, because these provide 'frame-works' for reviewing the event
- Learning: as a result of the review phase, consideration of what you might do differently next time will lead to new goals and actions. This might include deciding to definitely not try that again or alternatively to acknowl-edge that a strategy worked rather well and will be used again
- Apply: applying your new learning and going through the cycle again.

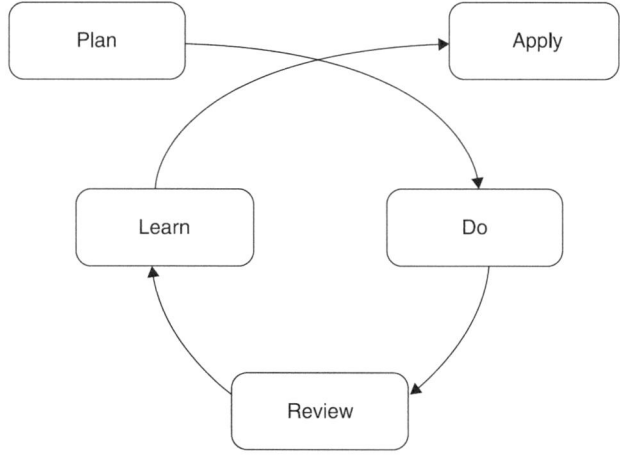

■ **Figure 2.1** A model of active learning

(*Source*: Watkins et al., 2007, p. 27)

Effective mentors will be able to support you in all the phases of the active learning cycle, through reflective conversations of the type described in the next section.

REFLECTIVE CONVERSATIONS

As part of the cycle described earlier, your ability to have productive reflective conversations with colleagues will help you make substantial progress in your own classroom practice.

Your ITE programme is likely to include at least two school placements, usually in two contrasting school environments. Your first few days in school will involve acclimatisation and familiarisation with this new environment. You are, however, likely to find yourself very quickly in a classroom, observing experienced teachers in action, making notes or keeping a record of what you observe. Until you have had some practical experience, it is difficult to extract the finer points of what you observe happening in the classroom so that you can apply them to your own practice. Students often imagine they are about to start teaching a very well behaved, motivated group of pupils, having observed them with their usual class teacher. On starting to teach the class themselves, the pupils prove to be unexpectedly difficult to manage. The subtle management techniques employed by the class teacher take some time to unpick: finding time to discuss your observations with the class teacher or your mentor can make a major contribution to your understanding of the complexities of successful teaching.

Some early questions you might ask the class teacher could include:

■ What strategies do you use to settle the class?
■ Do you time the different sections of the lesson?

During the feedback from a lesson observation, your mentor (or tutor or the class teacher) will expect you to reflect on the lesson. The first question from them should give you the opportunity to share this reflection, and it is usually 'How do you think the lesson went?' The natural response to this is something along the lines of 'Quite well' or 'Not very well'. While this may be a useful starting point, it is not very helpful for either of you! Consideration of the following points may help to focus your answers so that a constructive conversation can follow:

■ Comment on the different sections of the lesson. For instance, 'The pupils seemed well engaged with the 'do now' activity and it helped them settle quickly' or 'The starter went on for much too long and so they didn't have enough time to complete the main activity'.

■ If you have carried out a practical, there are usually multiple points you could reflect on. For instance, 'The movement of the pupils around the class was rather chaotic' or 'My explanation was clear enough for them to complete the practical safely and on time'.

■ If part of the lesson involved asking questions and expecting 'hands up' from the pupils, you can comment on how effective this was. For instance: 'the same few pupils answered all the questions' or 'I directed questions to specific pupils but some of them couldn't answer the questions'.

This brief list of suggested scenarios may provide an opportunity for your observer to agree or disagree with your reflections and to pose questions to you. Their purpose will be to help you think about whether specific strategies worked well and could be incorporated in later lessons or adapted for other classes or to revisit your plan and consider how to improve and/or adapt aspects for future lessons. Observers could ask the following sorts of questions:

■ What do you think is an appropriate length of time for a starter activity?

■ How could you change the starter to ensure it is an appropriate length of time?

■ How could you organise movement around the lab at the start of a practical to ensure safety and efficient management of apparatus?

■ Would you do anything to change your explanation of. . . ?

The list is of course endless, but the most important point for a good reflective conversation is that you have the opportunity to voice your reflections on the lesson and your mentor encourages you to think through alternatives and improvements. At some point they will make suggestions and work with you to identify targets for the next lesson. Your provider will have a system for you to use to keep a record of these reflections and targets. The importance of your contribution to these conversations cannot be underestimated. As you build your knowledge, expertise, and experience, your ability to critically self-assess and reflect is an essential part of your personal development as a teacher, and in this first year of teaching you have the opportunity to discuss your reflections with experienced colleagues. Starting to develop these skills as a beginning teacher will stand you in good stead for the whole of your teaching career.

ITE providers are likely to use particular models and approaches to support mentors and student teachers with observation of and feedback on lessons, which can mean these reflective conversations have a different tenor and tone. These might include guidance on observation and on how to set targets based on an observation that are not too generic but are appropriate and actionable by student teachers. One increasingly popular format of support for student teachers and, indeed, experienced teachers in schools, is known as *instructional coaching* (Knight, 2007). Unlike other coaching models, instructional coaching is rooted in the principle that there is a more expert coach supporting someone who is more of a novice in that particular area. The expert coach observes teaching to identify a particular area to focus on that they think will make the biggest difference to the teacher's practice. They then identify a target that might help them take a step towards achieving their overall goal of more impactful or effective teaching in that particular area, model and discuss a suggested strategy to achieve that target, observe the teacher practising that strategy and offering tightly focused feedback, and then, when that strategy is secure, work with the teacher to set another target to aid them in reaching their overall goal. Crucial to any form of useful feedback and target setting based on lesson observations is the relationship you have with the tutor or mentor who is working with you to offer that feedback, support, and coaching: this facilitates a quality reflective conversation regardless of the model used.

REFLECTIVE JOURNALS

Keeping a reflective journal regularly may seem like yet another requirement in a long list of tasks to complete as your practice develops, but students who do this frequently find their progress is enhanced. Before starting a journal, consider development of a format and style that suits you; jot your thoughts down immediately after a lesson, in notes or bullet points, and leave space to include feedback from your observer and to reflect on this feedback. Over time you may see themes or patterns appearing and you can consider how these may affect your learning and your teaching strategies. Be honest in your entries; your journal should be personal and private to you.

> It was really useful to write what you thought straight away as you might not get to talk to [the observer] until the end of the day . . . and then it would be interesting to see if they match up. To start with [your observations compared to your observer's observations] would be completely different . . . but towards the end it would match up much better.
>
> (Evans, 2021)

By keeping this sort of record, you will have excellent evidence of your progress towards developing as a reflective practitioner, as well as building your confidence in your capabilities to 'read' what is happening in your classroom.

Task 2.2 **Starting and keeping a reflective journal**

Choose a format that you will be able to access regularly: a small notebook you can carry with you if you prefer pen and paper or a tablet you keep with you during the day. The importance of accessibility is so you can capture thoughts and feelings in the moment, as well as reflecting on situations after they have occurred. Little and often is better than trying to do all your entries once or twice a week. Find a style that suits you – notes, bullet points, narrative writing, and graphic notation. It can be helpful to divide 'pages' into first stage reflection and second stage reflection by dividing the page in two and using the left-hand column for your immediate reflections. After a few days, go back over a number of pages and see where there are themes and patterns coming up in what has been written. What are you noticing? What might this mean for future learning and strategies? Capture your insights in the right-hand column.

THE ROLE OF METACOGNITION IN REFLECTION

During your ITE course, you are likely to learn about metacognition and how teachers can support pupils to self-regulate their learning and use metacognitive strategies to support their learning. The term 'metacognition' was originally coined by John Flavell in the late 1970s. He defined the word as 'cognition about cognitive phenomena' (Flavell, 1976) or basically higher-order thinking about thinking. The Education Endowment Foundation identified self-regulation and metacognition as one of their seven strands of effective science education (Holman & Yeomans, 2018). Alongside the utility of approaches that teach our pupils how to plan, monitor, and evaluate their learning, including modelling thinking and promoting metacognitive talk and dialogue (ibid., p. 14), metacognition is an important tool in your own learning to teach science. It involves understanding how your thinking is developing which in turn leads to better planning, delivery, and effectiveness of your teaching.

> [Metacognition] basically encompasses knowledge about one's own thought processes, self-regulation and monitoring what one is doing, why one is doing it and how what one is doing helps to solve the problem (or not).
>
> (Muijs & Reynolds, 2011, p. 155)

While this quote is referring to the use of metacognition from a pupil perspective, Artzt and Armour-Thomas (1998) neatly describe its applicability to teachers: 'the problem to be solved is how to teach a lesson that will promote student learning with understanding' (p. 6). They suggest that the components of meta-cognition that seem to impact on effective teaching are the thinking and reflection that occur during:

■ Planning before the lesson
■ Monitoring and regulating during the lesson
■ Assessing and revising after the lesson.

These activities will be an integral part of your practice, but the effectiveness of your constructive reflection at each point will depend on how self-aware, self-determined, and self-directed you are. Your metacognitive thinking should lead to both monitoring the impact of your teaching 'live' during the lesson and adjusting as a result and also reflecting on the lesson afterwards and deciding how you might adjust your teaching in the future, both when teaching the same topic again and when teaching other topics and classes.

In the case of a student science teacher, the 'problem' could be how to teach a particular topic most effectively and to be able to reflect on the success of the chosen strategy after the event: did the pupils learn what you intended them to learn and were they able to use their learning successfully, i.e. meet your success criteria?

The following scenario illustrates what good metacognition might look like:

> John is planning a lesson on electric circuits. His learning intention is for pupils to understand current and voltage in series and parallel circuits and his success criteria are that pupils will to be able to answer a series of set questions at the end of the lesson on current and voltage in circuits. By the end of the lesson his assessment (which could be via a series of verbal or written questions or simply by observing the students) suggests that the pupils were confused, particularly with questions about voltage. His initial reflections suggested that he over-ran on time and that his explanation of current and voltage was not clear enough for the pupils to have understood and be able to apply their knowledge to the questions. He decides that he must include less content in his lessons and clarify his explanations – these are the obvious outcomes of his reflection. As he now starts to think further about this he decides that he needs to use a better model to explain current and voltage (so perhaps he will have a conversation with the physics teacher about what models might be used) and he needs to script his explanation in advance of the lesson, not to read from the script but to clarify his own thinking and to have a logical narrative sequence when presenting his explanation. Perhaps he also needs to script some quite specific and sequential questions to encourage his pupils to think in a metacognitive way about the circuits.

The outcome of this for John's teaching is that he now has a better understanding of the difficulties for pupils of understanding abstract concepts in physics, of what he can hope to achieve in one lesson, and a way forward to improve his teaching.

Task 2.3 **Metacognitive reflection on an aspect of a lesson or part lesson you have taught**

Using the scenario described earlier as a guide, choose a lesson or part lesson to develop your metacognitive skills. Start by reflecting on the lesson (perhaps using your reflective journal to record your thoughts). Consider the choices you made in planning your lesson – strategies to deliver the objectives – including materials and resources – and to monitor learning. How and why did you come to make these choices? Reflect on the success of these, particularly in relation to pupil learning, using any evidence you feel appropriate. Now for the metacognitive bit: how will you adjust your thinking about these choices in your next lesson to facilitate better learning? This might involve researching and choosing a better model, asking more (or less) challenging questions, sequencing your questions so that they build thinking and learning more effectively, using better visualisation to illustrate key points, or improving the content and sequence of your explanations. The list of possibilities is endless, but the process of monitoring your original thoughts and adjusting for the future is at the heart of the exercise.

In their early lessons, student teachers typically have so much to think about, metacognitive monitoring and reactive adjustment during the lesson can be very challenging. In the first instance, you are more likely to think metacognitively about your teaching through a reflective process following the lesson. However, as you develop your skills, your capacity to use metacognitive strategies to adjust your teaching in response to pupil's learning during the lesson will help you become an excellent teacher.

EVALUATING PRACTICE

Reflection and evaluation of lessons are vital for a teacher as it is a way we learn and further develop. To learn from your plans you need to evaluate them after the event or to reflect 'on action' (Schön, 1983) and this is done in the process of lesson evaluation. Evaluation is not just about making a value judgement (saying something was good or bad or I liked it or did not like it), which then becomes a self-critical negative tool. It is far more effective to use it positively but is a very challenging thing to do. There are different models available to help do this but a basic reflection and evaluation (based loosely on Gibbs, 1988) includes the following components:

- ▪ *A description* of what happened in the lesson without any value judgement attached to those descriptions. This is an important step as it will be the evidence on which to base your evaluation. If possible, jot down notes on your plan as you go along, e.g. note down timings, anything you did differently or forgot to include, anything that didn't work, problems you encountered. Then after the lesson, note down anything that happened differently to what you expected.

■ *Evaluation*: Here you can judge (or get others to help you judge) what went well and what didn't go so well, what was good or bad about the lesson. Writing down your reflections immediately after the lesson, especially early on in your teaching practice, will encourage you to articulate your thoughts and give you a basis from which to have professional discussions with any observer. It will also give you a narrative to look back on: were your initial reflections based more on emotional responses to what might have been a difficult lesson and less on a critical self-awareness of your delivery and your pupils' learning?

■ *Analysis of the lesson*: Why did it go the way it did? Why did I get those outcomes? For example, if your timings were out, why was this? It is important to be brutally honest with yourself at this stage even if it is uncomfortable.

■ *A suggestion of further actions/action plan/self- improvement targets*: what could I do better/different next time? What can I learn from this experience? What actions do I need to take? Actions might include investigating new approaches, building up your subject knowledge, or attending training, as well as doing something differently in the next lesson. This is especially important for people who are working with you, as selfreflection – if it is done well – helps those people (mentors and tutors) to support you.

You might encounter lesson plan templates that include evaluation templates with them. An example of this is seen in Figure: 2.2, taken from the Exeter University 'Beginning Practice Phase' lesson plan proforma.

EVALUATION: Your Teaching

You may choose to evaluate this lesson using these boxes, and/or by annotating your lesson plan.

Please evaluate your lesson against the Profile Descriptor for the phase you are in. If you are using this template this is likely to be the profile descriptor for the Beginning Practice phase. Please analyse what went well or did not go well in your lesson and why. Brief bullet points.

The impact on pupils' learning: You may wish to select target pupil(s)/group(s) as the focus of your evaluation *You might like to consider how well the pupils:*

• *succeeded in meeting the learning objectives*
• *applied skills, knowledge or understanding to meet the lesson objective(s)*
• *engaged with the lesson*
• *used the resources available, including adult support, to improve their learning*
• *used self/peer assessment to improve their own learning.*

*In evaluating the lesson, indicate **how you know** that your teaching has had an impact on pupils' learning. Brief bullet points*

■ **Figure 2.2** Evaluation section taken from the University of Exeter 'Beginning Practice' lesson plan proforma, 2022

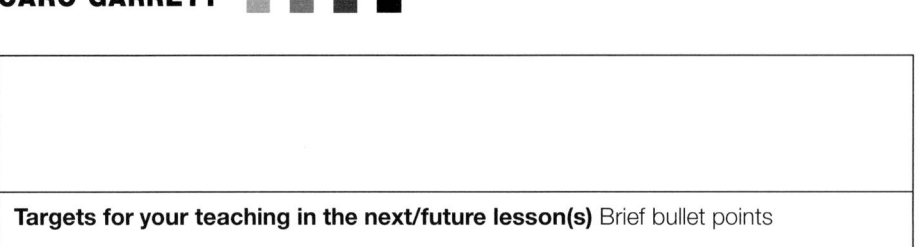

Targets for your teaching in the next/future lesson(s) Brief bullet points

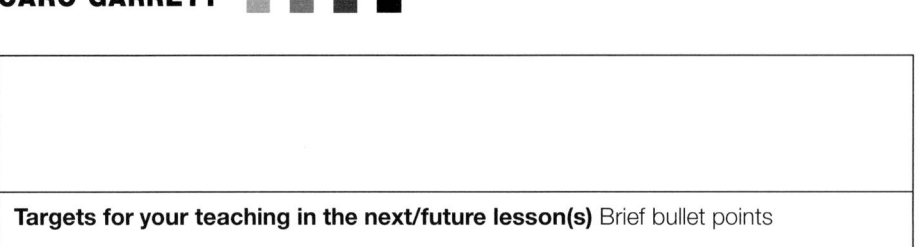

■ **Figure 2.2** Continued

Task 2.4 **Evaluate a lesson**

Use the template in Figure 2.2, or one supplied by your ITE provider, to evaluate a lesson. Make sure you focus on positives as well as negatives – this exercise is not about 'beating yourself up' but about building skills to be effective in your self-reflection and to use these reflections to improve your teaching and your pupils' learning. Early in your practice these reflections may seem mostly negative, but use your mentor and your class teachers to help you be realistic and constructive.

Part of your evaluation of a lesson will include your evaluation of the learning of your pupils. How you assess this is covered in Chapters 16 and 17, and following your assessment, you need to reflect on the effectiveness of your plan in assisting your pupils in their learning. This may be discussed in a reflective conversation with an observer of your lesson, but as you progress in your teaching journey, the opportunity for these discussions will diminish. As a student teacher, you will have time allocated during the school day for planning; one effective way of using this time is to discuss your plan with a fellow student teacher before the lesson and then to talk through your evaluation after the lesson. Articulating your thoughts informally to a 'critical friend' rather than a formal observer can be a productive learning experience.

REFLECTION AS A KEY VEHICLE FOR PROFESSIONAL DEVELOPMENT

As a student teacher you are at the start of a life-long learning journey as your teaching career progresses. During your first years you will be given extra opportunities to engage in professional learning activities, some of which will be directed, but you are likely to also have the opportunity to self-direct your development. You will already have some idea of gaps in your subject knowledge that need to be filled, and there will be an expectation that you will work independently to fill these gaps. However, once you have familiarised yourself with *what* your pupils need to learn, *how* you will teach it (your pedagogical knowledge) will give you much pause for thought. Formal Continuing Professional Development (CPD) will be made available to you and your colleagues through your provider during your ITE and your school throughout your career. (For teachers in England, the Early Career Framework (ECF) is a structured form of professional development that you are required to undertake in your first two years in post.) CPD is often focussed on generic topics such as behaviour management, pedagogical skills, or curriculum design. You should use your reflection on your own personal development needs to identify what courses or sessions will be of particular use to you; these may entail engagement with peers or colleagues with particular skills. Science specific professional development may be made available to you through your school's involvement with outside agencies and can be accessed through a variety of sources such as learned societies (Institute of Physics, Royal Society of Chemistry/Biology); professional associations (Association for Science Education, STEM Learning); and school, university, and private providers.

Task 2.5 **Exploring available continuing professional development (CPD)**

Research available CPD on the website of a Learned Society or Professional Association, using the previous list as a starting point. Other possible sources are sciencemuseum.org.uk, SAPS.org.uk (Science and Plants for Schools), earth-scienceeducation.com, and futurelearn.com. The UK government have also published 'A guide to STEM CPD opportunities for teachers' (DfE, 2016a; link in the references).

SUMMARY AND KEY POINTS

In this chapter we have looked at the importance of reflection as a means of improving your practice and developing your skills of monitoring, analysing, and developing your teaching before, during, and after lessons. We have looked at the role of reflective conversations with mentors and how as a student teacher you can help to participate in and enrich these conversations. The use of a journal has been encouraged, as has the importance of making time for good quality reflection, including evaluation. The place of metacognition in reflection and in contributing to making progress in your development has been explored, as has the importance of reflecting on your needs for – and accessing – professional development.

FURTHER RESOURCES

Capel, S., Leask, M., Younie, S., & Lawrence, J. (Eds.). (2022). *Learning to teach in the secondary school: A companion to school experience* (9th ed.). Routledge.

Chapter 5 includes useful information about the use of practitioner research and reflective practice to improve your teaching.

Education Endowment Foundation. (2018). *Improving secondary science: Guidance report*. Education Endowment Foundation.

This useful report, written by John Holman and Emily Yeomans, highlights seven key recommendations for improving science teaching and contains a useful section on self-regulation and metacognition.

Education Endowment Foundation. (2021). *Metacognition and self-regulation toolkit*. Retrieved October 2023, from https://educationendowmentfoundation.org.uk/education-evidence/guidance-reports/science-ks3-ks4

This toolkit for teachers outlines key findings and strategies for supporting the development of metacognition and self-regulated learning.

Gilbert, J. K. (2010). Supporting the development of effective science teachers. In J. Osborne & J. Dillon (Eds.), *Good practice in science teaching: What research has to say* (2nd ed.). Open University Press. Retrieved September 2021, from https://channay-ousif.files.wordpress.com/2011/06/good-practice-in-science-teaching-what-research-has-to-say.pdf

This chapter looks at why teacher development is important and presents a model to enable progression. It also discusses the range of themes that may be included in developmental activities and what is entailed in becoming a more effective science teacher at any phase of a professional career.

SAFETY IN SCIENCE TEACHING

Luke Graham and Jill Noakes

INTRODUCTION

The safety of children in our care is a primary responsibility of all teachers. It is no accident that the first of the Teacher's Standards opens with keeping pupils safe, stating that, as part of standard 1, teachers must 'establish a *safe* and stimulating environment for pupils, rooted in mutual respect' (DfE, 2011, p. 10, our italics). Whilst this standard refers, for all teachers, to the setting of high expectations of pupils and the need for promoting good behaviour, it is particularly important for science teachers who are working in laboratory settings where there are particular additional and obvious hazards above other classroom teaching. This chapter focuses on the responsibilities of teachers of science for pupil safety in both practical and non-practical settings, including maintaining up-to-date knowledge of safety legislation, school and departmental policies, and the assessment and management of risk. This will enable you to confidently include interesting and exciting practical activities in your teaching to help promote pupil learning, whilst maintaining a safe learning environment. The chapter then exemplifies how these ideas might work in practice using two examples, to show the thought process experienced teachers use when planning for safe practical learning activities.

OBJECTIVES

At the end of this chapter, you should be able to:

- Understand the responsibilities of a teacher or new teacher in science
- Explore the ways teachers keep pupils safe in practical and non-practical sessions
- Reflect on the risk assessment process and how it applies to your practice.

DOI: 10.4324/9781003110187-4

There are many examples in teaching where teachers can disagree on the most appropriate or efficient way of delivering a topic. Should you start with an open-ended task or with some teacher led, direct instruction content? Have pupils working individually, in pairs, or in groups? Draw on a practical activity, or not? In these matters, you can make a good case for any approach. You can make the decision based on your knowledge of the pupils in the room, the time of the lesson, their relationship with you, the topic you are teaching and, ultimately, what you believe will be the best approach to help them learn the particular objective. Teachers are professionals and although they will work within some constraints based on school policies, they are all different and will make different professional decisions in the best interests of their pupils to drive the learning. Teachers can decide on a lesson-by-lesson basis what is best for the class. As a teacher, you can decide how you want the class to work and how you would like your pupils to develop.

Health and safely is not like this. There is clear legislation that must be followed, and a designated person in the school will have created a policy that you must know, understand, and follow. This designated person is known as the 'competent person' and must be someone with the necessary knowledge, skills, and experience to give sensible guidance about managing the health and safety risks as the school (DfE, 2022). It is likely that the overall competent person for a school will have delegated the management of health and safety risks associated with the science department to the Head of Science. There should be clear policies and practices that are explained to you as a new member of science teaching staff (student teacher or employed by the school), and it is your responsibility to diligently follow these arrangements and ask for specific help or training if you are not confident. Schools would always much rather you ask if you are not sure of whether to do something or how to do it safely. Accidents are very rare in school because teachers are diligent.

There are two general rules you should always follow:

1 You must take care of your own health and safety and that of others who may be affected by your actions. This requires thinking in advance about problems and not just reacting when they arise.
2 You must cooperate with your employer, obey any codes imposed, and follow the health and safety arrangements in your school.

In the next sections of this chapter, we will delve deeper into these. We will explore teachers' duty of care to pupils and the legislation in place that secures this. We then go on to explain some typical approaches to the management of health and safety concerns in science in secondary schools and where to find more information, before looking at how to use this information to assess the risks associated with common hazards in school science. Finally, we will exemplify how these processes work using teacher examples.

Teachers' duty of care to pupils

Teachers are required to do all that is reasonable to protect the health, safety and welfare of pupils. Under current legislation, these responsibilities derive from three sources: the common law duty of care, the statutory duty of care, and the duty arising from a contract of employment. It should be noted that as a teacher, you will have a responsibility to maintain your awareness of changes to health and safety legislation, and you will be supported in this by your school and the competent person. In common law the standard of care expected of each teacher is the skill and care of a 'reasonable teacher'. In other words, the question to ask yourself – or that could be asked of you should an accident happen – is 'did you do what any other reasonable teacher would have done?' The common law duty of care is influenced in part by the age of the pupils and what you are teaching them.

Rooted in legislation, teachers have additional responsibility for the pupils in our care, and it is from this that the statutory duty of care arises. For example, the Children Act requires teachers to have regard for the need to safeguard and promote the welfare of children when carrying out their work (The Children Act 2004; Section 11). The Children and Social Work Act 2017 amended these acts and includes provision regarding local safeguarding practice and relationships and sex education. There are various other statutory documents relating to child protection and pupil safety, and the one most prominent in schools is Keeping Children Safe in Education (DfE, 2023) which is statutory guidance for anyone working with children in England.

Finally, the duty of care arising from a contract of employment will make additional requirements upon you to protect the pupils in the school. Different schools will have different contractual obligations, but typically these derive from the 'school teachers' pay and conditions document' (STPCD). These will include statements like: teachers must 'promote the safety and well-being of pupils and staff'. As a student teacher in a school there will be a document between you and the training provider and the school that covers similar content.

These wider issues related to the safeguarding of children will be covered thoroughly on all teacher education programmes, and you will regularly be required to update your training in this respect as a student teacher and then as a teacher employed in school. For the rest of this chapter, we will focus specifically on issues of health and safety related to school science laboratories.

Accidents in school are very rare

Schools are required to report serious accidents, outbreak of disease, or dangerous incidents to the Health and Safety Executive and injury rates are low. It is difficult to find a recent detailed breakdown of reported statistics on injuries caused by accidents in science laboratories, but the ASE Health and Safety group suggest that the conclusions of dated research into accident rates continue to be valid, as on the whole practices have not significantly changed. This research (Tawney, 1981; validated by ASE in 2016) suggests that 4 to 5 per cent of pupil accidents in secondary schools occur in laboratories. Most of these injuries were caused by chemicals on the body, followed by cuts, burns, and scalds; dropping, falling, slipping, knocking, and lifting; chemicals in the mouth; inhalation; animal bites; explosions; fainting; and electric shock.

While accidents are rare they do need recording, no matter how minor. Schools are required to keep a record of all accidents at their premises and it is your responsibility to follow your school policy and record appropriately if you or any pupils in your class sustain any injuries in your lessons. Anyone injured at work is required to tell the employer and record details in an accident book, including the answer to the question 'how did the accident happen?' The employer is required to investigate and enter this in the accident book if they find anything that differs from the entry made by the worker. These regulations are intended to ensure a record is available in case there is a claim for compensation.

Your school will have both a competent person (likely to be called the Health and Safety officer) and an accident book or a way of adding accidents to a central register. As part of your induction into the school you should be made aware of both. The school (or employer) has a responsibility to provide staff with 'adequate health and safety training' (Regulation 13 of the Management of Health and Safety at Work Regulations 1999). Paragraph 2(a) of the regulation makes clear that employers must provide induction training for new staff. Although such training is the responsibility of the employer, at least part of this training is likely to be done by the head of science or some similar manager within the science department. 'Induction' here refers to what 'new members of staff need to know on a day-to-day basis in order to teach the curriculum without significant risk to the health and safety of themselves or others (pupils, technicians, cleaners, classroom assistants, etc.)' (CLEAPSS, 2017). Part of this training will include specific information about how to complete and record risk assessments within your school policy.

Risk assessment in school

CLEAPSS (2022) define a **risk assessment** as 'a judgement of how likely it is that someone (anyone) might come to harm if a planned action is carried out'. Risk assessments are legally required for actions carried out at work, with significant findings recorded. Essentially, in conducting a risk assessment as a science teacher, you need to:

1 Conduct an analysis of a practical activity to identify **hazards** and the steps needed to minimise the risk of harm occurring
2 Record the conclusions of your analysis.

A hazard is anything that can cause harm. Risk assessment involves identifying both the likelihood that the hazard would cause harm and the extent of that harm in terms of both the severity of potential injury and the number of people it could affect. **Control measures** are put in place to minimise risk of harm, such as the wearing of protective personal equipment (PPE) such as safety glasses or gloves.

The first purpose of the risk assessment is for the responsible adults to reflect on the procedures being employed, think about what is going to happen, and work to reduce the potential harm. It is not a process for preventing enthusiastic teachers from creating learning experiences that are new or unusual (see Association for Science Education, 2020). The written risk assessment document is not a risk assessment; it is a summary of the risk assessment process. There is no point in finding a risk assessment that has been previously produced and recorded in the school and filing it without further thought. The thought process is the most important part. Most school activities have school-wide measures in place to deal with the common risks, so that teachers and support staff do not need to produce written assessments for an ordinary classroom unless new activities lead to additional risks. In science, it is likely that most standard practical activities will have a risk assessment already prepared and written into any scheme of learning that is being used or, alternatively, access to risk assessments produced by a national service such as CLEAPSS. These are helpful, but you should always read them carefully before conducting a practical and consider whether they might need to be adapted for the needs of the particular class you are teaching in the particular lab you are teaching in. Any adaptations to the standard risk assessment could be recorded on your lesson plan, in a planner, or in another location as determined by your school policy (which you should, of course, always follow).

Finally, it is important to remember that risk assessment is never a static activity that is 'completed' when conclusions are written down. Risk assessment is dynamic and may need to be changed as conditions change. This might mean abandoning a practical activity if something occurs during the lesson that changes your assessment of the risk, such as a particular behaviour exhibited by a pupil or group of pupils as they come into the classroom or a change in weather conditions if outdoors.

Task 3.1 **School policy on accident and risk**

For your current school, find out the school policy regarding a) how risk assessments are conducted, submitted, and reviewed and b) how accidents are recorded.

Task 3.2 **Risk assessing a lesson**

All science teachers have routines embedded in their lessons. These are processes and behaviours that the pupils do automatically. Observe a science practical activity in your setting. Make a note of the assumed procedures that the pupils adopt (e.g. bags and coats, chairs, desk organisation, where they move and in what order). Make a list of the direct instructions that the teacher gives related to the conduct of the practical activity (e.g. wearing safety glasses, lab coats, collecting and returning equipment, any safety instructions). Next, find a published risk assessment for the activity you have observed from CLEAPSS or your school documentation and compare your notes from the lesson with that risk assessment. Consider how the teacher might a) have adapted the risk assessment for that particular class in advance and why and b) whether you think the teacher changed anything during the lesson as a result of a dynamic risk assessment and why.

Thinking about assessing risks and benefits

The main questions teachers ask themselves when planning to include a practical activity are likely to be:

1 Is this the best way to teach this scientific concept?
2 If yes, then what are the hazards associated with this practical?
3 What can I do that will reduce the risk? (Examples might include tying hair back, standing up to conduct practical work, wearing safety glasses.)
4 What things might happen that I need to know the procedure for? (Examples might include a fire for which you need to know where a fire extinguisher or sand bucket is and how to use it, or where the gas stop is, or where the eye wash is and how to use it correctly to wash out a chemical splash in an eye.)
5 What adjustments do I need to make for *these* pupils in *this* class doing *this* practical next week? (Examples might include adjustments for pupils with an individual education plan, such as someone who faints at the sight of blood or who is partially sighted.)

The most common form of accident in a lab is a chemical spill, and there is little reason to think this might have changed over time (Tawney, 1981; validated by ASE in 2016). It is therefore essential that all teachers know what to do if the chemicals they are using get on the pupils' skin or eyes. Do they need to just rinse it? Where is the eye bath? How long should the eye be washed? Are you going to accompany the pupil to the eye bath? What are you going to do with the rest of the class? Who is available/on call?

All these questions (and more) should be pre-considered, and you should know the answer to them before you start the practical. There is insufficient space in a chapter such as this to go through the varied hazards associated with standard science practicals. The best resource we recommend for you to use to identify hazards, assess risks, and know how to mitigate them are the CLEAPSS resources, which are regularly updated. Your training provider and school will likely both be members of CLEAPSS and will be able to provide access to these resources.

In the final sections of this chapter, we will look at two examples to contextualise the information we have provided so far and to offer further insight into the thought processes you will undertake as a science teacher. These are written in the first person, as examples from experienced teachers. Please note that these are not instructions for how to conduct the practical or full risk assessments and should not be treated as such. For detailed risk assessments, please refer to CLEAPSS or your school schemes of learning or standard risk assessments. Rather, the intention here is to offer you insights into the thought processes these teachers have undertaken.

Practical examples

Biology: heart dissection practical walk through (Luke Graham)

There is no requirement to do a dissection in biology. Some of my colleagues do them and some don't. I do a demo heart dissection with my 11-14 year old groups and my 14-18 year old classes do a heart dissection in groups themselves.[1] The limiting factor is usually the cost of the hearts, which schools usually source from a local butcher. As the material is from a butcher it is akin to cutting up a steak and therefore safe to dissect. You will notice that organs will often have been cut into at the abattoir (as part of the inspection process) and come with slices in them.

I always get hearts from different animals, most of the hearts come from pigs, but I get some lambs' hearts too to accommodate various religious preferences. I also give the pupils an opportunity to do an online dissection rather than do one live. There are always some pupils who don't want to see it at all, and I have seen some pupils faint. So now I make it clear that there are three ways to participate, that the pupils can choose, and that they can change their minds during the practical. They can do the online version; or they can participate in a group; or they can watch through a screen (I set up a flex camera and put it on the screen). This is an adjustment I have introduced to help pupils participate, while recognising that the smell can be overpowering, and some pupils are more likely to faint than others. Before the session I make sure the windows are open to help clear the smell. Some pupils find seeing the action through a screen helps them, as it gives them a sense of remoteness.

In my mind I have a hierarchy of hazards. That a scalpel might go missing and be used in school is high up there. That someone might faint and hit their head. The risk of contamination by animal material. That a student might cut themselves while doing the practical, all the way down to concerns about the waste material ending up in the wrong bin, or pupils getting very little out of the practical and not learning much that lesson if they are unable to see or participate.

With these concerns in mind I ask the lab techs to prepare a tray for each group. It contains a clearly labelled (pig/lamb) heart, paper towels, and some straws. The hearts are on top of a paper towel bed and covered over. Pupils have access to gloves.

Before the pupils do any sort of incision, I get them to use the straws to identify where the vessels go to and leave from. To have a really good look at the structure and see if they can identify the front and the back. To see the non-symmetrical shape and to carefully feel (using gloves) the relative thickness of the two sides of the heart. This is important as pupils who know what they are doing and what they are looking at are more likely to participate in the practical in the way I intend. Once I am confident that they have got these main ideas, then I will hand out either scissors or numbered scalpels. You can do the practical perfectly well with scissors, but I think using a scalpel gives you more control and yields better results. If I decide that this group will use scalpels, then I allocated each one to a named pupil and write their name on the board next to the number that they have taken. At the end of the practical that pupil will return the scalpel to me (in its case) and I can rub off the name. Each scalpel has a new blade and is in its own case so it can be carried safely across the room to the tray where the heart is kept at all times. Scalpels are only out of the case when then are being used. These are my mitigations for my class in my school, and I have thought about the hazards and how to limit the risks. It is this thinking that is the crucial part of the risk assessment.

In terms of the risks associated with the use of scalpels there is some excellent material on the CLEAPS website (see resources section later).

Task 3.3 **What is the educational value of heart dissection?**

Search online for videos of virtual heart dissections and watch two. Search online for a paper on this topic or refer to one of those listed in the resources section below. Please bear in mind that papers from international sources such as the USA are common, but the use of dissection will be different in different jurisdictions. Having watched the videos and read the paper, reflect on your position on the use of animal material in biology lessons. Can you offer a rationale for your position?

Physics: using ray boxes for optics experiments (Jill Noakes)

It is not immediately obvious that using small lamps in ray boxes to study the behaviour of light carries much risk. There are the standard electrical safety considerations of course; using a low-voltage power supply requires the same basic care as any other electrical equipment. However, there are two safety aspects to this practical that require not only pupil awareness but some careful pedagogical consideration.

The most significant thing about optics practicals is that they have to be done in the dark. Preferably very dark, although classrooms vary in terms of how practical this is and the availability of blackout blinds. I therefore need to ensure that the classroom is even safer than usual in terms of general manoeuvrability; can we get around the room without tripping over or crashing into things? Bags need to be tucked away, chairs pushed in and all walkways generally kept clear. I need to be especially aware if there are any pupils in my class with visual impairments; it may be helpful to discuss it with them and position them in a lighter part of the room or near the front so they don't need to move very far to leave the room if they need to.

Any concerns about what pupils might get up to are amplified in the dark, so I might consider compromising by making do with semi-darkness if I have concerns about behaviour. Working in the dark also makes it difficult to use visual cues and non-verbal communication, so calm, clear verbal instructions are more important than ever, as my usual stern looks and gestures for behaviour management are unlikely to be effective in these circumstances.

It is always important to make sure instructions for any practical are crystal clear before pupils start work, but this is even more true when you are about to turn off the lights. They will not appreciate me interrupting their work and dazzling their eyes by turning the lights on again because there is something I forgot to show them or because I need to clarify the instructions. However, turning the lights on *will* be an effective way to get the whole class's attention if this is necessary on behaviour management or safety grounds.

Some newer models of ray box use LED bulbs and therefore do not generate much heat, but the standard type used in most schools are filament bulbs that get hot when in use. The whole metal casing of the ray box gets hot, and the longer it is on for, the hotter it gets. The obvious advice might be to turn the ray box off when it is not directly in use; however it is also the pupils' only light source during the experiment, so they will typically leave it switched on while writing down results and any other work they may be doing while waiting for the class practical to end and for the lights to go back on. I am therefore going to make sure they are aware of this risk, turn it off whenever possible, and take care when handling it, even after the lamp has been switched off.

Having read these examples, it is useful to consider how the teachers are using their own judgement to assess the risk and consider why they would include these activities and how they might mitigate any risks, in line with their own personal teaching approach.

Task 3.4 **Analysing a risk assessment**

First, having read these two examples, reflect on how the teachers are assessing risk and the sorts of things they are thinking about. Refer back to the five points at the start of this section. Next, go to the CLEAPSS website (see further resources) or your school scheme of learning or risk assessment file and look up the standard risk assessment for one of these practical activities. In the teacher's descriptions, what is similar, and what is different? What do you think might be the reasons for these differences? What do you think you might do, if you were to conduct this activity with one of your own classes? Write up a risk assessment of your own.

As highlighted at the start of this chapter, teachers have professional agency in making decisions about their own teaching, in line with school policies with due regard to relevant statutory duties. As you develop your teaching career, risk assessment will be an ongoing key aspect of your role. Hazards and risks are a part of life, and certainly a part of science teaching and learning. Knowing how to assess risk and find information about mitigation will enable you to more confidently engage your students in learning through practical work safely and ideally enrich both your teaching and their learning.

SUMMARY AND KEY POINTS

Now you have read this chapter, you should have a basic understanding of:

■   Your duty of care as a teacher to safeguard pupils
■   Some of the essential legislation that underpins this duty and the impor-
    tance of school policies with respect to health and safety
■   Your responsibilities specifically in relation to health and safety in the man-
    agement of practical work, including reporting accidents and assessing and
    mitigating risk
■   What a risk assessment involves and how to find, adapt, or produce risk
    assessments for the specific classes you teach
■   How to include dynamic assessment of risk in your planning and teaching.

Check which requirements for your initial teacher education you have addressed through this chapter.

NOTE

1   Currently, the rules in Northern Ireland for dissections are slightly different to those
    in England, Wales, and Scotland

FURTHER RESOURCES

Association for Science Education (ASE). (2020). *Safeguards in the school laboratory*
    (12th ed.). ASE is a comprehensive guide to working in a school laboratory.
    CLEAPPS produces a set of 'HazCards' which can be found here https://science.
    cleapss.org.uk/resources/hazcards/. This is a substance-by-substance reference

for using and dealing with each of the substances you are likely to use in the lab. You should read the HazCard for the materials you intend to use. Alongside these, CLEAPSS' Practical Procedures pages which can be accessed here https://science. cleapss.org.uk/resources/practical-procedures/all/?search=px are very helpful to look at alongside the HazCards.

The Royal Society of Chemistry has an interesting article on the urban myths of banned practicals (https://www.rsc.org/images/Surely_thats_banned_report_tcm18-41416.pdf) which should be referred to if you are considering whether a practical can be safely executed.

There are two key organisations for science teachers; one is CLEAPSS (https://www. cleapss.org.uk) the other is ASE (https://www.ase.org.uk). Both have been referenced in the chapter, and we recommend their material as a good starting point. For example, the CLEAPSS guidance regarding dissection here: Retrieved September 22, 2023, from http://science.cleapss.org.uk/Resource/G267-Dissection-a-starter-guide-to-health-and-safety.pdf is a must-read when starting to plan a dissection in a lesson.

THE NATURE OF SCIENCE

Michael J. Reiss

INTRODUCTION

In the UK, most university students who study science are taught little explicitly about the nature of science. And yet the science National Curriculum in England requires pupils in schools to be taught about 'working scientifically' and similar ideas are included with the science curricula in Northern Ireland, Scotland, and Wales, not to mention many other countries. Perhaps unsurprisingly, research evidence suggests that most pupils leave school with a somewhat limited knowledge of this area of science. Yet, this can be one of the most satisfying parts of the science curriculum to teach. If you succeed in getting your pupils to have a reasonable understanding of how science is undertaken, they will be in a much better position to examine critically the many claims about science that they will come across in their lives – claims about disease, biodiversity loss, energy supplies, pollution, 'dangerous chemicals', and much more.

This chapter explains what is meant by the terms 'working scientifically' and 'the nature of science'. It looks at whether science always proceeds by the objective and rigorous testing of hypotheses or whether there are other factors at play in deciding whether one scientific view comes to hold sway within the scientific community over alternative views.

Within schools, there are different views about the nature of science among both pupils and science teachers. These different views are important in terms of how people see scientific knowledge. Ways of determining people's views on the nature of science are given later. It is hoped that these will be of interest and lead to richer teaching and learning in this area. This is important as it can be argued that long after pupils have forgotten much of the content of science that they are taught at schools, they will still hold a view as to how science is done and as to whether scientific knowledge is trustworthy or not.

DOI: 10.4324/9781003110187-5

OBJECTIVES

At the end of this chapter, you should be able to:

■ Distinguish between alternative understandings of how science is carried out
■ Contrast science as undertaken by scientists and science as undertaken in school science lessons
■ Help your pupils develop a deeper understanding of the nature of science.

WHAT DO WE MEAN BY HOW SCIENCE WORKS AND THE NATURE OF SCIENCE?

In the science National Curriculum, 'working scientifically' is described under four headings:

■ Scientific attitudes
■ Experimental skills and investigations
■ Analysis and evaluation
■ Measurement.

This covers a lot of ground but encapsulates quite well what is meant by 'the nature of science'. At both Key Stage 3 and Key Stage 4, the science National Curriculum goes into some detail about each of these four headings. This detail will not be presented here – it is found in almost all Key Stage 3 and Key Stage 4 pupil textbooks. Rather, what I will do is go beyond what pupils are expected to know, so that you will be in a better position to understand the underlying issues when teaching pupils and dealing with any questions they ask.

WHAT DO SCIENTISTS STUDY?

It is difficult to come up with a definitive answer to the question 'What do scientists study?'. Certain things clearly fall under the domain of science – the nature of electricity, the arrangement of atoms into molecules and human physiology, to give three examples. However, what about the origin of the universe, the behaviour of people in society, decisions about whether we should build nuclear power plants or go for wind power, the appreciation of music and the nature of love, for example? Do these fall under the domain of science? A small number of scientists would argue 'yes' to all of these and the term 'scientism' is used, pejoratively, to refer to the view that science can provide sufficient explanations for everything.

However, most people hold that science is but one form of knowledge and that other forms of knowledge complement science. This way of thinking means that the origin of the universe is also a philosophical or even a religious question – or simply unknowable; the behaviour of people in society requires knowledge of the social sciences (e.g. psychology and sociology) rather than only of the natural sciences; whether we should go for nuclear or wind power is partly a scientific issue but also requires an understanding of economics, risk, and politics; the appreciation of music and the nature of love, while clearly having something to do with our perceptual apparatuses and our evolutionary history, cannot entirely be reduced to science.

While historians tell us that what scientists study changes over time, there are reasonable consistencies:

- Science is concerned with the natural world and with certain elements of the manufactured world – so that, for example, the laws of gravity apply as much to artificial satellites as they do to planets and stars.
- Science is concerned with how things are rather than with how they should be. So, there is a science of gunpowder and in vitro fertilisation without science telling us whether warfare and test-tube births are good or bad.

HOW IS SCIENCE DONE?

If it is difficult to come up with a definitive answer to the question 'What do scientists study?', it is even more difficult to come up with a clear-cut answer to the question 'How is science done?' Indeed, there is – and has been for many decades – active disagreement on this matter among academic historians, philosophers, and sociologists of science. A useful point from which to start is with the views of Robert Merton and Karl Popper. In my experience, most working scientists have almost no interest in and know little about the philosophy and sociology of science. However, they do like the views of Merton and Popper once these are explained to them – though they generally think their arguments so obvious as not to need stating. The same working scientists are a great deal less keen on the views of Thomas Kuhn and others, to which we shall come in due course.

Robert Merton characterised science as open-minded, universalist, disinterested and communal (Merton, 1973). For Merton, science is a group activity: even though certain scientists work on their own, all scientists contribute to a single body of knowledge accepted by the community of scientists. There are certain parallels here with art, literature, and music. After all, Cèzanne, Gaugin, and van Gogh all contributed to post-impressionism. But while it makes no sense to try to combine their paintings (the very notion is absurd), science *is* largely about combining the contributions of many different scientists to produce an overall coherent model of one aspect of reality. In this sense, science is disinterested; in *this* sense it is (or should be) impersonal.

Of course, individual scientists are passionate about their work and often slow to accept that their cherished ideas are wrong. But science itself is not persuaded by such partiality. While there may be controversy about whether the works of J. S. Bach or Mozart are the greater (and the question is pretty meaningless anyway), time (almost) invariably shows which of two alternative scientific theories is nearer the truth. For this reason, while scientists need to retain 'open mindedness', always being prepared to change their views in the light of new evidence or better explanatory theories, scientific knowledge grows (though not uniformly) over time. As a result, while some scientific knowledge ('frontier science') is contentious, much scientific knowledge ('core science') can confidently be relied on: it is relatively certain. Think, for example, about the advent of COVID-19. It was a new virus so there was much in the early months that was poorly known – such as how infectious it was, how dangerous it was, and whether one would develop immunity if one had become infected and survived. This was all frontier science. But scientists have been studying viral diseases for decades. There was much core knowledge that could be drawn on – such as the fact that antibiotics would not destroy the virus and that in the absence of a vaccine the key way to control the pandemic would be ensure that each infected person on average infected fewer than one person.

Karl Popper emphasised the falsifiability of scientific theories (Popper, 1934/1972). Unless you can imagine collecting data that would allow you to refute a theory or hypothesis, the theory/hypothesis isn't scientific. The same applies to scientific statements. So, the statement 'All swans are white' is scientific because we can imagine finding a bird that is clearly a swan (in terms of its appearance and behaviour) but is not white. Indeed, this is precisely what happened when early white explorers returned from Australia with tales of black swans (Figure 4.1).

■ **Figure 4.1 Black swans (cygnus atratus)** are native to Australia. Until they became known to people in the West, the statement 'All swans are white' was thought to be true. However, the statement is falsifiable (capable of being falsified). Now we believe that this statement is false. Karl Popper argued that for a theory, hypothesis, or statement to be scientific, it needs to be falsifiable.

Popper's ideas can give rise to a view of science in which knowledge steadily accumulates over time as new theories are proposed and new data collected to discriminate between conflicting theories. Much school experimentation in science is Popperian: we see a rainbow and hypothesise that white light is split up into light of different colours as it is refracted through a transparent medium (water droplets); we test this by attempting to refract white light through a glass prism; we find the same colours of the rainbow are produced and our hypothesis is confirmed. Until some new evidence causes it to be falsified (refuted), we accept the hypothesis.

There is much of value in the work of Thomas Merton and Karl Popper but most academics in the field would argue that there is more to the nature of science. We turn to the work of Thomas Kuhn (Kuhn, 1970). Thomas Kuhn made a number of seminal contributions but he is most remembered nowadays by his argument that while the Popperian account of science holds well during periods of *normal science* when a single paradigm holds sway, such as the Ptolemaic model of the structure of the solar system (in which the Earth is at the centre) or the Newtonian understanding of motion and gravity, it breaks down when a scientific *crisis* occurs. At the time of such a crisis, a scientific revolution happens during which a new paradigm, such as the Copernican model of the structure of the solar system or Einstein's theory of relativity, begins to replace the previously accepted paradigm. The central point is that the change of allegiance from scientists believing in one paradigm to their believing in another cannot, Kuhn argues, be fully explained by the Popperian account of falsifiability.

Kuhn likens the switch from one paradigm to another to a gestalt switch (when we suddenly see something in a new way) or even a religious conversion. As Alan Chalmers puts it:

> There will be no purely logical argument that demonstrates the superiority of one paradigm over another and that thereby compels a rational scientist to make the change. One reason why no such demonstration is possible is the fact that a variety of factors are involved in a scientist's judgment of the merits of a scientific theory. An individual scientist's decision will depend on the priority he or she gives to the various factors. The factors will include such things as simplicity, the connection with some pressing social need, the ability to solve some specified kind of problem, and so on. Thus one scientist might be attracted to the Copernican theory because of the simplicity of certain mathematical features of it. Another might be attracted to it because in it there is the possibility of calendar reform. A third might have been deterred from adopting the Copernican theory because of an involvement with terrestrial mechanics and an awareness of the problems that the Copernican theory posed for it.
>
> (Chalmers, 2013, pp. 107-108)

Kuhn also argued that scientific knowledge is validated by its acceptance in a community of scientists. Often scientists change their views as new evidence persuades them that a previously held theory is wrong. But sometimes they cling obstinately to cherished theories (scientists are human). In such cases, if these scientists are powerful (e.g. by controlling which articles get published in the most prestigious journals), scientific progress may be impeded – until the scientists in question change their mind, retire, or die!

In a chapter of this length there clearly isn't space to provide an account of many other views of how science is done but there is one philosopher of science whom you will either love or hate – Paul Feyerabend. In many ways Feyerabend anticipated the postmodernists with their suspicion of a single authoritative account of reality. His views are succinctly summed up in the title of his most famous book *Against Method* (Feyerabend, 1993). Feyerabend is something of an intellectual anarchist (another of his books is called *Farewell to Reason*) and the best way to understand him is for you to read him rather than for me to provide too tidy a summary of his thinking. Here are a few quotations – to either whet your appetite or put you off for life:

> No theory ever agrees with all the facts in its domain, yet it is not always the theory that is to blame. Facts are constituted by older ideologies, and a clash between facts and theories may be proof of progress.
>
> (Feyerabend, 1993, p. 39)

> the events, procedures and results that constitute the sciences have no common structure;
>
> (Feyerabend, 1993, p. 1)

> the success of 'science' cannot be used as an argument for treating as yet unsolved problems in a standardized way.
>
> (Feyerabend, 1993, p. 2)

> *there can be many different kinds of science*. People starting from different social backgrounds will approach the world in different ways and learn different things about it.
>
> (Feyerabend, 1993, pp. 2-3)

It should not be thought from these quotations that Feyerabend was sceptical about the value of science. Quite the opposite – he was passionate about science and its worth. What he didn't like was people, especially if they weren't scientists themselves, laying down the law as to how good science is undertaken. The best scientists are rather like the best artists – they use every means at their disposal to tackle a problem. School science is great at providing an introduction to how scientists work, but it's just that – an introduction. It's rather like teaching poetry to a group of 12 year-olds. You might tell them about rhyme and scansion, leaving other forms of verse till later.

Task 4.1 **Getting pupils to consider the scope of science**

It is all too easy for pupils to take for granted that they get taught different 'things' in their different subjects in school. The aim of this task is to help pupils to become more aware of why there are certain things they study in science lessons as opposed to in other lessons.

You can start by having a discussion with pupils about whether there are some things in science that they also learn about in other subjects (e.g. rocks in geography, muscles and breathing in PE, sound in music). Then get them, perhaps in pairs or small groups, to talk about the similarities and differences between what they get taught about in, for instance, geography and what they get taught about in science. Pupils should end up appreciating that there are overlaps but that science is more interested in experiments, is more universal (so that most science is much the same in any country whereas geography is very interested in differences between countries) and has less to say about human action unless such action can be studied objectively.

Discuss the findings with your mentor and explore how any cross-curricular links are built in your school.

Task 4.2 **What is your view of the nature of science?**

There have been a number of research instruments devised to enable a person's views on the nature of science to be determined. Probably the easiest for you to get hold of and use is that provided by two UK science educators – Mick Nott and Jerry Wellington (Nott et al., 1993). If you quite enjoy reading questionnaires along the lines of 'Does your partner find you boring?' in the backs of magazines while at the dentist, this task is for you. Once you have completed this questionnaire, reflect on *why* the scoring system came up with the result it did. Do you feel content with your view of the nature of science? How might your nature of science profile affect how you teach science? How would your answers have needed to differ for you to have been classified differently? Discuss whether there is a single ideal view of the nature of science that science teachers should hold.

'REAL SCIENCE' COMPARED TO SCHOOL SCIENCE

There is much in the 5 to 16 science National Curriculum in England with which to be pleased. However, one of the less successful areas has been what in the 2014 version is called 'Experimental skills and investigations'. For one thing, we don't do a very good job of getting pupils in school science lessons to ask the sorts of questions that scientists actually ask or to ask the sorts of questions that the public asks (e.g. 'Are vaccinations safe?', 'What happened before the big bang?', 'Should we all go vegetarian?') and to which science can make a contribution. Instead, pupils are too often restricted to uninteresting questions about the bouncing of balls or the dissolving of sugar, in what are misleadingly termed 'scientific investigations'.

The history of this part of the science curriculum since the first National Curriculum Science Working Group, which published its report in 1987, up to the year 2000 has been analysed by Jim Donnelly (2001) and remains relevant to this day. Donnelly argues convincingly that two conflicting understandings of the nature of science can be detected in the battles of those years (and they felt much like battles to those participating in them). One is what Donnelly terms 'essentially empiricist'; in other words, straightforwardly concerned with how factual data are collected with which to test hypotheses. The other stresses social and cultural influences on science. In the language of the first part of this unit, this is a contest between Popper and Kuhn (Feyerabend doesn't get a look in).

The reasons for this battle need not greatly concern us except insofar as the very existence of the battle indicates the lack of consensus in this area among those responsible for the science National Curriculum. This contrasts with many other parts of the science National Curriculum. There is, for example, little controversy over the inclusion of food webs, evaporation, and series circuits.

Much school science operates on the assumption that 'real science' only consists in doing replicated laboratory experiments to test hypotheses. When I started teaching social biology to 16–19 year-olds in England, the Examinations Board (it was in the mid-80s, before Awarding Bodies) that set the syllabus included a project of their own design that pupils undertook. One of my pupils was a very fit athlete who lived in Bahrain out of term time. He undertook measurements on himself of such physiological variables as body temperatures and body mass just before and just after completing a number of runs both in the UK and in Bahrain under very different environmental conditions. Another pupil was interested to see whether a person's astrological sign (determined by their birth date) correlated with their choice of school subjects at advanced level (e.g. sciences versus arts) or with their personalities. Accordingly, she carried out quite a large survey of about 100 of her fellow pupils and ran lots of correlation tests.

Both pupils had their projects scored at 0 per cent. Nor were these marks changed on appeal. I'm not claiming that either of these projects was the finest I have ever seen. But I am convinced that the marks they were given reflected too narrow an assumption in the examiners' minds about what constituted a valid piece of scientific enquiry (Reiss, 1993).

Those who write science curricula and mark pupils' work are powerful determinants of what passes in school for 'real science'. A more valid approach to finding out what science actually consists of is to study what real scientists do. Careful ethnographic work (spending hours in laboratories watching the behaviour of scientists, studying their lab books or field notes, etc.) on this only began in the 1970s – one classic book is Latour and Woolgar (1979). Much of this writing is rather difficult to read and it's much easier to read any really good biographical account of a scientist – though such biographies, in concentrating on individuals, do sometimes give rise to the 'Great scientists' view of scientific progress, underplaying the extent to which science is a social enterprise, one that relies on the contributions of many, many individuals.

The take home message from the ethnographic work on science is that while scientific practice is partly characterised by the Mertonian and Popperian norms discussed earlier, there is plenty of support for Kuhn's views and even for Feyerabend's. For example, school accounts of the science often underplay the political realities of scientific research (not to mention the monotony of much of it). Governments and companies are far more interested in funding botanical research on the genetic modification of crops than on moss reproduction. Actually, it's the exception that proves the rule. If your moss only lives in Antarctica, chances are you may indeed obtain funding as many countries are keen to undertake research there so as to stake a territorial claim should the wilderness ever be exploited for natural resources or military purposes.

Task 4.3 **Ask pupils to research how science is done**

Get your pupils to research, using books in the school library or (more likely nowadays) internet sources, one example of the history and practice of science. For example, you might get pupils to research:

■ The theory of evolution
■ Rosalind Franklin's contribution to determining the structure of DNA
■ In vitro fertilisation
■ Developing a vaccine for COVID-19
■ The Periodic Table
■ The history of glass
■ Plate tectonics
■ The competition between DC and AC among domestic suppliers of electricity
■ The life and work of Galileo
■ The use of X-rays.

This exercise works best if pupils have a choice on what to work on and have some opportunities to work collaboratively. You might get them to think about such things as:

■ Were these scientists working on their own or with others?
■ How did they build on what was already known about the issue they were working on?
■ Did they collect new data or provide new interpretations of data?
■ How were their ideas received when they proposed them?
■ What similarities are there between the science they did and the science that is done in school?
■ What differences are there between the science they did and the science that is done in school?

If relevant, file some of the outputs in your personal development portfolio.

PUPILS' VIEWS OF THE NATURE OF SCIENCE

In the UK, a classic piece of research was carried out on pupils' views on the nature of science from 1991 to 1993, i.e. in the early years of the National Curriculum. The work was undertaken with 9, 12, and 16-year-olds in England and came up with the following findings (Driver et al., 1996):

- Pupils tend to see the purpose of science as providing solutions to technical problems rather than providing more powerful explanations.
- Pupils rarely appreciate that scientific explanations can involve postulating models. Even when they do, models are presumed to map onto events in the world in an unproblematic manner.
- Pupils rarely see science as a social enterprise. Scientists are seen as individuals working in isolation.
- Pupils have little awareness of the ways that society influences decisions about research agendas. The most common view is that scientists, through their personal altruism, choose to work on particular problems of concern to society.

One technique that has been used to examine how pupils see science is to ask them to draw a scientist (Figure 4.2). The drawings are then examined to see

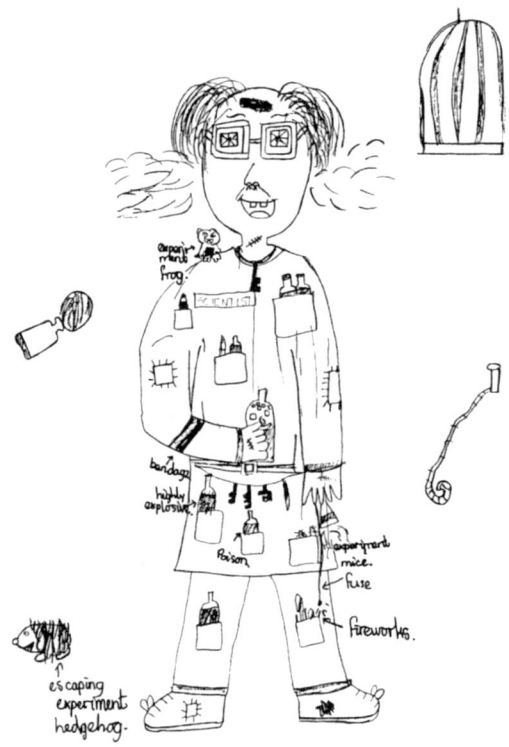

■ **Figure 4.2** Caroline Allen's drawing of a scientist

whether pupils tend to draw scientists as male or female, white or black, in laboratories or in other settings, etc. In the 1960s and 70s only about 1 per cent of girls drew a woman when asked to draw a scientist; in the 1980s about a third did; nowadays about half do (Miller et al., 2018). Substantial progress has been made – partly as a result of the efforts of teachers, partly as a result of changes in society more generally.

Task 4.4 **The 'draw a scientist' test**

Get your pupils to do the 'draw a scientist test'. (This probably works better at KS3 than at KS4.) Explain to them before they start their drawings that this isn't a test in the sense of some people gaining more marks than others and that you aren't interested in the artistic quality of their drawings. Ensure they write their names on the back of the drawing (in case you want to analyse the results by gender, ethnicity, science capital (see Chapter 18), performance in science, or something else that requires knowledge at the individual level).

Usually, researchers get pupils to do this individually but you might decide that it would be interesting to get pupils to discuss in pairs what they are going to put in their drawings before they do them.

If you want to go beyond what most researchers do, you might do the following and discuss your findings with your mentor:

■ Require pupils to write a few paragraphs explaining why they drew what they drew
■ Interview pupils about their drawings
■ Explore with pupils what they have seen or read that informed what they drew.

SUMMARY AND KEY POINTS

Science is concerned with the natural world, with certain elements of the manufactured world, and with how things are rather than with how they should be.

The nature of science is a controversial area and there are several competing understandings of how scientific knowledge is produced. Robert Merton saw science as open-minded, universalist, disinterested, and communal. For Karl Popper, the distinctive thing about science is that all its ideas are testable and so falsifiable. Thomas Kuhn went beyond these views in seeing certain crucial episodes in the history of science as inexplicable within Mertonian and Popperian frameworks of thought. Kuhn emphasised that when scientists switch from one way of seeing the world to another, i.e. as the paradigm changes, they do so for a variety of reasons, not all of which are scientific in the narrow sense of the term.

Pupils generally have rather a limited understanding of the nature of science. The tasks presented here are intended to help deepen their understanding.

Check which requirements for your initial teacher education course you have addressed through this chapter.

FURTHER RESOURCES

Chalmers, A. F. (2013). *What is this thing called Science?* (4th ed.). Open University Press.

> An extremely clear and very widely read introductory textbook on the philosophy of science. If you want a readable account about any of the major questions on the nature of science, here is an excellent place to look.

Kind, V., & Kind, P. M. (2008). *Teaching secondary: How science works*. Hodder Education.

> A really well written and intelligent overview of how science works at both Key Stages 3 and 4.

Kuhn, T. S. (1970). *The structure of scientific revolutions* (2nd ed.). University of Chicago Press.

> A classic in the history and philosophy of science. This is the book which undermined the straightforward Popperian view of science and convincingly argued for the importance of culture in the growth of scientific knowledge. Kuhn introduced the notion of 'paradigms' in science.

McComas, W. (Ed.). (2020). *Nature of science in science instruction: Rationales and strategies*. Springer.

> A very thorough guide to all aspects of the nature of science, edited by one of the acknowledged experts in the field. There are many chapters, written by different authors, with such titles as 'Beyond experiments: exploring valid and reliable methods in science', 'Exchanging the myth of the scientific method for a more authentic description of science' and 'Perspectives for teaching about how science works'.

What is the nature of science?

> https://scienceonline.tki.org.nz/Nature-of-science/What-is-the-Nature-of-Science. A helpful website for science teachers on what we want pupils to understand about the nature of science.

5

TEACHING BIOLOGY
Luke Graham

INTRODUCTION

The theoretical physicist Brian Greene describes the battle between two forces in the universe, entropy and evolution.

> The 2nd law of thermodynamics dictates the entropy of the entire universe, as an isolated system, will always increase over time. This means that matter tends to fall apart. But on the other side of the battle are the evolutionary forces of selection that tends to bring matter together into interesting structures that yield complex molecules and living and thinking systems. . . . The entropic two-step and the evolutionary forces of selection enrich the pathway from order to disorder with prodigious structure, but whether stars or black holes, planets or people, molecules or atoms, things ultimately fall apart.
>
> (Greene, 2020)

Until that great fall apart happens, there will be living, responsive organic systems, and thinking systems like us will study them, categorise them, and examine how they work and where they came from. Biology is the study of life and living organisms, including their physical structure, chemical processes, molecular interactions, physiological mechanisms, development, and evolution. But it is more than that: unlike physics or chemistry, it is the study of living systems – and sometimes living systems might not want to be studied.

Biology provides an opportunity to develop debate about key issues that are important to young people and to equip them with the skills and habits to be informed about the life choices they make (DeWitt & Archer, 2015). It helps them know about themselves a little bit more but might also get them to think about personal and social considerations such as hereditary questions, health and disease, and resource management.

DOI: 10.4324/9781003110187-6

This chapter looks at how biology is taught in the secondary classroom and the underlying biological principles we would like our pupils to understand and the scientific habits they should adopt.

OBJECTIVES

At the end of this chapter you should be able to:

- Understand how some key concepts in Biology could be taught
- Be aware of the features of living things
- Make links between different biological concepts
- Identify some common areas of misconceptions.

OVERVIEW OF THE BIOLOGY CURRICULUM

In Biology at Key Stage 3 (KS3), typically studied by pupils in England between the ages of 11 and 14, pupils are introduced to key scientific knowledge, conceptual understanding, and ways of working scientifically. It is more than the acquisition of biological facts, it is also a way of encouraging pupils to be scientists, rather than simply learn biology.

Themes through secondary biology pick up on six broad but (fairly) uncontentious scientific ideas.

- The use of conceptual models and theories to make sense of the observed diversity of natural phenomena (See Vosniadou, 2012.)
- The assumption that every effect has one or more cause
- That change is driven by interactions between different objects and systems
- That many such interactions occur over a distance and over time
- That science progresses through a cycle of hypothesis, practical experimentation, observation, theory development, and review
- That quantitative analysis is a central element both of many theories and of scientific methods of inquiry.

Key topics included in Secondary Biology at Key Stages 3 (age 11-14) and 4 (age 14-16, where pupils in England typically study GCSE examinations or equivalent) are shown in Table 5.1

It is not possible to cover every aspect of the five years of biology teaching here, so this chapter will focus on a few examples from some of these topics. These have been selected because they provide opportunity to explore how teachers can make the learning and understanding of biology accessible and because they offer insights into key misconnections or preconceptions (Clement, Brown, & Zietsman, 1989) that pupils have about the topic or the language around the topic.

■ **Table 5.1** Example national curriculum coverage (England)

	Working scientifically	Subject content
Ages 11–14 Key Stage 3	1 Scientific attitudes – making predictions, being accurate, using evidence, asking questions 2 Analysis and evaluation – Calculations, tables, graphs, explanations 3 Measurement – use of units	Structure and function of living organisms Material cycles and energy Interactions and interdependencies (ecosystems) Genetics and evolution
Ages 14–16 Key Stage 4	1 The development of scientific thinking 2 Experimental skills and strategies 3 Analysis and evaluation applying the cycle of collecting, presenting, and analysing data, including: 4 Vocabulary, units, symbols, and nomenclature	Cell biology Transport systems Health, disease, and the development of medicines Coordination and control Photosynthesis Ecosystems Evolution, inheritance, and variation

Key questions

As we go through the sections there are some key questions to keep in your mind.

■ How can I make this topic engaging? (DfE, 2014)
■ What are the links to the pupils' life experiences? (Skinner, Kindermann, & Furrer, 2009)
■ How can I make pupils' experience of science outside school valuable? (Archer et al., 2015a)
■ How can I link this topic to previous topics or to prior experience? (Brod, Werkle-Bergner, & Shing, 2013)
■ How can I help pupils understand by reducing unnecessary load on working memory (Sweller, 2011) and being precise with my language (Kane & Engle, 2002)?
■ How can I help pupils to enact or practice this new knowledge? To apply it to new situations, to different representations of it (Wirebring et al., 2015)?

CELL BIOLOGY

'Cells' is a common starting point for many schools in biology, perhaps because cells are so ubiquitous as 'cells are the building blocks of life' (from Schwann (1839) cell theory). It also gives pupils a chance to use (possibly for the first time) a stock instrument of the biologist: the microscope. Etymology (the breaking down of words into their components) can be a helpful way for pupils to lower the cognitive load of learning and using new words. What words start with micro? What does 'micro' mean?, what words end with scope? – telescope, periscope. Microscope – a thing for looking at small things.

The cells topic draws on material that the pupils have learned before (if they covered science at primary school to this level) and starts discussions about accuracy, validity, and resolution that will continue right though to biology A level and beyond. Cell drawing and microscope use are key skills that are picked up again and again though biology at school and commonly appear in GCSE as a core practical.

However, microscopes are expensive and not able to withstand being dropped and some models can easily drive the objective lens through the stage if you are not careful. Many schools have limited numbers of microscopes and you can have 3 or 4 pupils sharing one microscope. Most pupils will struggle to focus it, and many will ask the teacher to help them. Classes can occur where 30 pupils have drawn intricate pictures of air bubbles, the slide mount, the microscope stage, or mascara on the lens. It's an exercise in logistics and resource management as much as it is teaching how to draw a specimen accurately.

Most students will not have used a microscope before. Those that have are likely to have used a simplified one. School microscopes are usually different from each other as they are bought one or two at a time, as they are too expensive to buy in class sets. Some have lights, some have mirrors, some have condensers, and some don't. Some have holders for the slides where you wind them around, some you clip the slide under two (unless one is broken) metal clips and push it about with your finger, some have no clips at all. Be sure you know which sort of microscopes you have in school and you have used them before you unleash them on the pupils!

Task 5.1 **Microscope lesson**

Ensure you have access to a typical school microscope (you may need to do this task on placement) and make up a slide with a very small piece of paper from a magazine or newspaper (see Figure 5.1) and a spot of water (you will not need a cover slip). Look at this at low power (e.g. x4 objective lens) and medium power (x10) and high power (x40) if you have the lens.

Now consider, before the pupils even start drawing their specimen:

■ What are the barriers to pupils using microscopes?
■ What key information do pupils need the first time they use a microscope?

The following ideas are some suggestions but this is by no means comprehensive. Once you have thought for yourself, read through these ideas and reflect on the similarities and differences with your own ideas.

When I first get out the microscopes there are two parallel thoughts in my mind.

Point 1: How can I make sure this is safe and help the pupils conduct the practical without injuring themselves or breaking the microscopes?

How to carry a microscope. How to focus it with the 2 wheels without breaking the slides. Should I remove the high power (expensive) lens from the microscopes the first time they use them? How do I want them returned? Where should the pupils sit? Are they making a slide or being given a slide? Are they collecting and returning the microscopes? Are they sharing? Who is working with who? What do I need to demonstrate, what can they learn though experience?

Point 2: How can I ensure that the pupils take away key ideas about cells or microscope use and that the practical is engaging and purposeful?

How can I break the learning into smaller, manageable pieces? How long do the pupils need to have access to the microscope to see the image? Will they be able to make good progress unaided? Can I get round the class without being constantly asked to focus up their microscopes? Should they work on the microscopes in pairs while the other half of the class does a different task and then swap? Should they photo their cell/slide and then draw from the image so I can get more pupils though the observation part of the lesson? Should, I project up an image on the board so they can see what they are aiming for?

And I don't want my solution to point 1 to mean I spend 40 minutes demonstrating so they lose interest and forget how to do it by the time they start and thus undermine point 2.

When I teach this lesson for the first time I separate how to use a microscope from drawing the cells (Rosenshine, 2012). It does mean it takes more teaching time in the short term, but over the year you get that time back as the pupils are more efficient at collecting, using and returning the microscopes, so can focus on the slides we are looking at.

(Luke Graham)

Figures 5.1–5.3 show three images of the same slide at three different magnifications. (Etymology of the word magnification: *Magni* from *Magnus* – great or large like magnificent or a magnum ice-cream (or champagne), and *fication* from to make or do, like beautification or mummification.) The slide is made up of a small piece of magazine. It is deliberately chosen as my first slide to use with my year 7s as it is familiar. When they are looking down the eyepiece, they have an idea what they should be able to see. That way they are more likely to be able to focus on the object and create a clear image. Once they have it focussed, they might start to notice some things about the way microscopes work as well as noticing interesting things about the image and the letter on the magazine. Biologists notice things. Biologists say, 'hmm, that's interesting, I wonder why that happens?' a lot.

Noticing things is important, and something pupils need time to practice.

They should notice that they can see more detail at higher magnification, the paper is not smooth, it is made up of fibres, the letter is not clearly defined nor a constant tone or colour at medium power. They should also notice that the

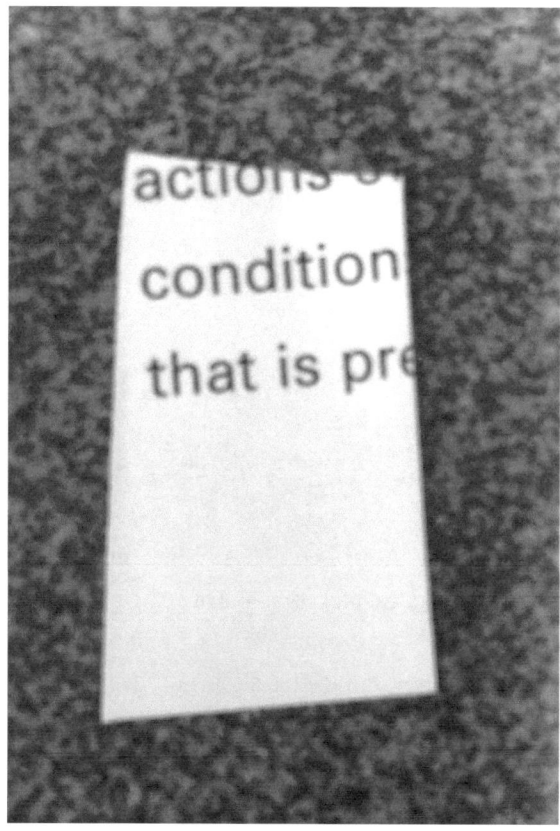

■ **Figure 5.1** A piece of printed text photographed at actual size

■ **Figure 5.2** A piece of printed text photographed at 40x magnification

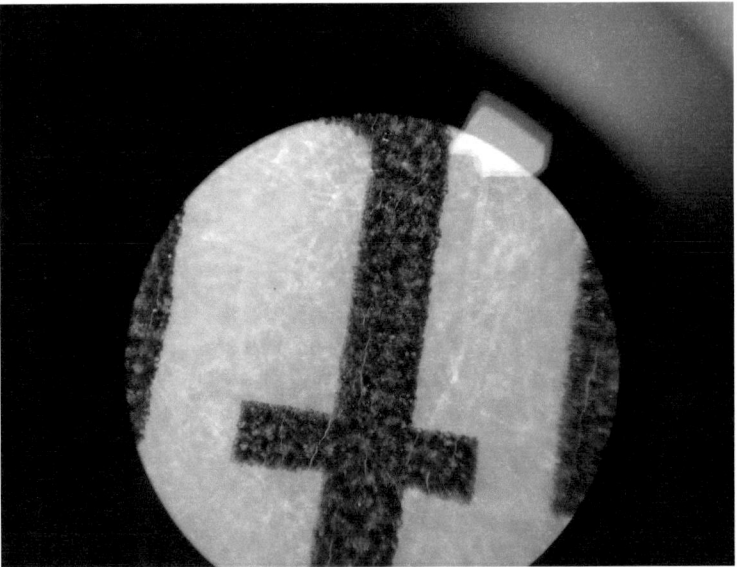

■ **Figure 5.3** A piece of printed text photographed at 100x magnification

image is inverted, and if they move the slide left the image moves right, and the image moves quicker than the slide. They might also notice that at higher magnification, they need to use the smaller/fine focus wheel. They might even notice that the water is needed to stick the paper to the slide so it does not curl up or blow away or move about.

All of these observations will make it easier when they need to look at the next slide later in their science lessons.

Task 5.2 **What might pupils notice?**

Consider what the pupils might notice when looking at print through the microscope for the first time, drawing on the previous ideas and any others you might add to the list. Now, think about the question: If a pupil was to ask you **why** one of the things they have noticed is happening, what answer would you give? Assume this is for an early secondary lesson, and try to make multiple links (Brod et al., 2013) in your answer to other primary and secondary topics in biology or in other sciences (such as lenses/light).

Cells and organelles

As well as being able to identify different parts of cells in a microscope or from an image or drawing, we would like pupils to know what those parts of cells do. Hattie and Yates (2013) summarise the ways humans collect and organise knowledge such as 'ideas' or propositions about the world around them. For example,

'*Some plant cells contain photosynthetic organelles called chloroplasts*' or '*Animal cells have mitochondria*'

   Here conceptual meanings are linked. For example:

Chloroplast: *Chloro* means green [like chlorophyl or chlorine] and *plast* like plasticine means made of.

Mitochondria: *Mito* is thread like and *condria* means like a grain so here the etymology may not be as useful as 'thread like grain' is not a great definition of what it is or does. But thread like grain does represent what it looked like to early scientists with limited power microscopes!

We can test that knowledge by asking pupils questions such as, 'Do animal cells have mitochondria?' This gives us insight into what the pupil knows or does not know. But it does not tell us *why* they don't know the answer, or what we can do to help them understand. Are they unfamiliar with the word 'cell' or 'mitochondria' for example? Did they learn mitochondria and chloroplast in the same lesson and have confused their meaning? We can teach pupils that the appropriate answer to the question 'do animal cells have mitochondria' is 'yes'. But this is superficial and inefficient learning. The amount of new information you can hold like this is limited (Rosenshine, 2012; Sweller, 2011). The way that teachers use questioning and structure tasks is extremely important in helping teach in ways

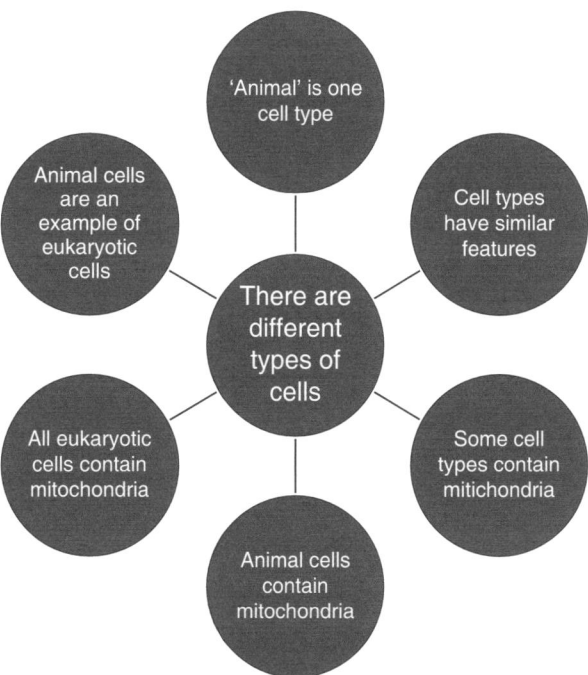

■ **Figure 5.4** Concepts about cells

■ **Table 5.2** A simple table of the features of plant and animal cells

	Animal cell	Plant cell
Cell wall	Absent	Present
Shape	Round (irregular shape)	Rectangular (fixed shape)
Vacuole	Absent	Large vacuole
Chloroplast	Absent	Present
Cytoplasm	Present	Present
Mitochondria	Present	Present

that aid learning and encourage pupils to remember key ideas. Take this typical simplified table of animal and plant cells for example (Table 5.2).

The issue with this black and white approach is that it is overly simplistic: It encourages students to think in strings of data, not patterns, concepts, or ideas (Hattie & Yates, 2013), and also it is only sometimes true, not always true. For example, where would an onion cell, or a phloem cell, or a mature red blood cell fit in this table?

A key question for a teacher is, 'how can we help pupils take these ideas and make them into more organised, robust and efficient schema?' (Hattie & Yates, 2013). One way could be to replace the 'Present/Absent' with 'always/sometimes/never'. Animal cells **never** contain chloroplasts, but a plant cell **sometimes** contains them. We could also ask the pupils to give an example if it was a 'sometimes' answer. Can they give an example of a plant cell that does contain chloroplasts and one that does not? A small adjustment to the table used offers a richer seam of assessment gold to mine. It also encourages pupils to move away from thinking things are one thing or another, (it is photosynthesising **or** it's respiring, or a bee is not an insect it's an animal for example (Allen, 2014)), rather than it could be both at the same time.

A Venn diagram might help to contextualise the similarities and differences and get pupils thinking harder about what the key feature of cells are. By asking the pupils to group, sort, or organise we can help them to arrange related items into a meaningful pattern (Hattie & Yates, 2013, p. 123).

Task 5.3 **Cells Venn diagram**

Add the following cells to a Venn diagram like the one shown in Figure 5.5:

Onion cell	Phloem cell	Seaweed
human skin (adipose)	heart (cardiac) muscle	mature red blood cells
lung alveoli cell	fish gill (plate) cell	A virus e.g. HIV
white blood cell		A bacteria like E coli

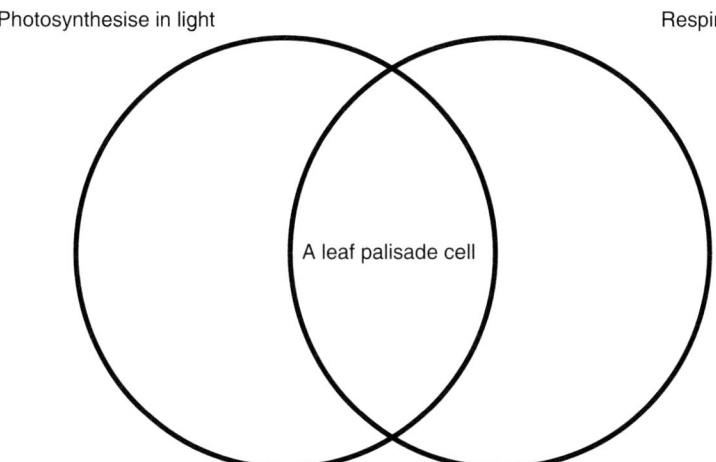

Photosynthesise in light   Respires all the time

A leaf palisade cell

■ **Figure 5.5** A Venn diagram of cells that photosynthesise and/or respire for you to complete

Now consider these points

- Assume that the previous exercise is aimed at KS4 pupils
- How can you adapt it for KS3? How would you adapt this exercise to make it a 'low threshold, high ceiling' task (NRICH, 2011)?
- Is the list of cell types in a sensible order?
- Are the last few cells suitably challenging at this level, or too challenging? What can you add to improve this exercise?
- Does the label 'respires all the time' rather than just 'respires' help in this exercise?
- What biological terms might be missing from the labels or the cell examples?

Using the information in an unfamiliar context might help pupils to make this learning more permanent (Wirebring et al., 2015). For example, what-if questions might dig deeper into pupils' understanding of biology. For example: *Cyanide stops the mitochondria working. What would happen if I added cyanide to an animal cell? What if I added it to a plant cell?* This might lead pupils to point out that it would kill the animal outright, but the plant cell might be fine in the day. A follow up question to probe this answer would then be helpful, asking '*Why would it be ok in the daytime?*'. The pupil may answer that plants respire at night and photosynthesise in the day. This cyanide example shows how this if-then question exchange identifies a common misconception about plants (Maeng & Gonczi, 2019), meaning the teacher can then do something to reconstruct this pre- (or mis-) conception (Allen, 2014).

Another example of a 'what-if' question is, '*What might happen if I covered a leaf so there was no light?*' (Note how using *might* here rather than *would* changes the question?) This question might provide us with a similar opportunity to unpick the 'respiration at night' misconception. Or *'What might happen if I put a plant in a room where there was only one hour of light each day?'* Initially, both questions might get the answer 'it would die', but *why* will it die? What is the biological reason for that? In answering these questions, pupils have *described* what would happen, but the really important question teachers would need to probe is whether or not they can *explain* why it would die. The key terms here are 'describe' and 'explain'. These are 'command words' that give important instructions to pupils about what it is that they are being asked to put into words.

As they are learning, pupils construct organised schemas of knowledge (see Chapter 10) and careful questioning, such as suggested here with respect to some common misconceptions in the cells topic in biology, allows teachers to assess pupils' learning and prior understanding. Asking pupils to organise and access knowledge across different formats gives them an opportunity to think more deeply about what we want them to be thinking about.

Task 5.4 **Command words and teacher questioning**

Have a look at the exam board guidance on command words linked to in the 'Further Resources' section at the end of this chapter; these are the words and phrases that tell students how they should answer a question across all exams.

When you are in school make a note of the sorts of questions teachers ask the pupils. Are they *name* questions? Or are they *describe* questions or *explain* questions? Do teachers use these command words explicitly or others with similar meanings? How do teachers test and probe deeper understanding? For example, 'but is that always the case?' or 'can you give me an example?'

Can you make any links between how the pupils respond to the question and the way the question is phrased? How long do the pupils get to think about each sort of question in class?

Where next?

You might want to explore 'Cells, organelles and cell processes' and how the outputs of the mitochondria are used by other cell organelles. You might want to start with or move on to the characteristics of life or onto the evolution of multicellular organisms, the evolutionary story behind the two key infections in cell biology: The infection *or endosymbiosis* of cells by mitochondria about 1.45 bn years ago and the infection by chloroplasts about 1 bn years ago were two evolutionary events that set in motion the evolution of plant and animal life on the planet. *Endo* means 'within', like endocytosis and endoscope, *sym* means 'together', like sympathy, and *biosis* means 'a way of living' There is a great chapter in *Other Minds* by Godfrey-Smith (2016), in which he explains this story in evolutionary history very well.

However, we are going to move on to how the metabolites are provided to cells and tissues in multicellular animals (including humans) and how the waste products are taken away. In exploring this topic, we will continue to think about the importance of command words and vocabulary in teacher questioning and in how the learning is structured and sequenced. In primary science the term 'animals including humans' has replaced the former 'animals and humans'. Why is this I wonder?

TRANSPORT SYSTEMS

What is your favourite blood vessel?

Everyone should have a favourite.

Mine is the Hepatic Portal Vein, because it is the only vein that travels between two organs and carries a varying quantity of glucose in healthy humans. It would be understandable if you went for the classic pulmonary vein or the sensitive aorta instead.

'Transport systems' is another classic topic in biology, running from primary science right though A level and beyond. The topic includes the structure or arteries, veins and capillaries, the names of the vessels and the organs they serve, and the chambers of the heart and the passage of oxygenated and deoxygenated blood. There is a lot of subject content in this unit, including many new and unfamiliar terms like *semi-lunar* and *renal* and *endothelium* - or *epithelium* and *systole* and *hepatic* and *ventricle*? This language can confuse pupils and it is important in teaching this topic to consider the literacy demand on pupils.

Task 5.5 **The hepatic portal vein**

With the previous information, can you:

Name the two organs the **hepatic portal vein** connects?

Describe the role of the two organs connected by the HPV in the homeostasis of glucose.

Explain why the quantity of glucose varies in the **hepatic portal vein** and not in other blood vessels.

Suggest why the organ that the hepatic portal vein travels to needs an additional artery to supply it.

For pupils aged 14–16 (KS4 in England), what links are there from 'transport systems' to other parts of the biology scheme of work? For example, to cellular respiration?

Transport systems is one of the topics that really lends itself to thinking about the big picture before you dive into blood vessel differences and system names. If we understand the **function** of the transport system, then the **structure** is much more accessible. This link between **structure** and **function** is another classic biological refrain. How is the **structure** of the lungs adapted to its **function**? Or the palisade layer in a plant? Or the retina (which is an interesting example because the answer is 'not very well because it is limited by the evolutionary pathway'!).

So, let us begin with the big picture. What is the big picture description of the mammalian transport system? What does it do? What is its function? If we ask this question to any collection of pupils and scientists, we would get a range of different answers:

- It pumps blood round the body
- It takes oxygen from the lungs to the other organs
- It transports glucose, oxygen and other metabolites to the cells and takes the waste/excretion products away
- It is for homeostasis like maintaining body temperature, pH
- It allows organisms to be larger. Small organisms can diffuse the metabolites they need though their external surfaces. Large organisms cannot, so they need a way of getting the external components inside
- It gets the immune cells to the site of any infection.

These are all correct answers at some level. Some lack detail and could be improved with some deeper questioning (Paul & Elder, 2007, 2019), but the key point is there are many functions of blood and the transport system, and in any one lesson you are only going to cover some of them. However, we need to be clear that blood can do many jobs at the same time, (just like a bee is *both* animal and insect).

Having established the big picture, we can move on to explore the finer detail. We might explore questions such as, 'What is the difference between arteries and veins?' Here are some typical answers to this question:

- Arteries carry oxygenated blood and veins carry deoxygenated blood
- Veins have higher glucose concentration than arteries
- Veins have valves and arteries do not
- Veins take blood to the heart and arteries take blood away from the heart
- Arteries carry blood away from the heart towards an organ, while veins carry blood from an organ towards the heart.
- Arteries have higher pressure blood than veins
- Arteries carry blue blood and veins carry red blood

Looking at the previous statements, it is important to consider which statements are only *sometimes* true and which ones are *always* true or *never* true.

Which ones are consequences of other answers – for example, is the presence of valves a consequence of traveling 'upwards' to the heart? Or a consequence of low pressure? If it is it a consequence of low pressure, why are there valves in the heart where the pressure is the highest? Unpicking answers to these sorts of questions and relating the big picture with the fine detail are important in planning for learning. Biology teachers need to ask themselves, for this topic and others, 'do we want to start with the function or the structure? Does one make more sense, does one lend itself to build on the previous learning more sensibly, or do we need to do both at the same time?'

There is not usually a right or wrong answer to these questions, as what a teacher chooses to do in organising the learning depends on many factors. It depends on the group, on the topic, on the order of the topics in the scheme. Most important, however, is that it is an deliberate choice, not an accidental one, and you can articulate the reason in your approach to the lesson planning.

SUMMARY AND KEY POINTS

This chapter considered some important ideas about subject content knowledge (SCK) and pedagogical content knowledge (PCK; Berry & Loughran, 2010; Loughran, Mulhall, & Berry, 2004) drawing on examples from the curriculum. We looked at the ways teachers can support learning and challenge pupils to think more deeply. We examined the use of the assessment judgements we make to improve the learning of our pupils.

Check which requirements for your initial teacher education you have addressed through this chapter.

FURTHER RESOURCES

DfE (Department for Education). *The National Curriculum for in England: Science programme of study*. https://www.gov.uk/government/publications/national-curriculum-in-england-science-programmes-of-study/national-curriculum-in-england-science-programmes-of-study

It is useful to look carefully at what is included in the curriculum for your country or jurisdiction, to see the basic sequence of content and key ideas. The one for England is linked to here, covering content from age 4 to 16.

In the UK, exam boards who set end-point examinations such as GCSEs and A-levels produce a document called 'command words' and they are used in very much the same way across all subjects and boards. It is useful for you to be aware of what these are. Here are links to AQA and Pearson.

Godfrey-Smith, P. (2016). *Other minds: The octopus and the evolution of intelligent life*. William Collins.

Goldacre, B. (2010). *Bad science: Quacks, hacks, and big pharma flacks*. McClelland & Stewart.

Harari, Y. N. (2014). *Sapiens: A brief history of humankind*. Random House.

These 'popular science' books are a great read. We use all three as prizes for our pupils. They are accessible and full of interesting nuggets of gold to sprinkle in your lessons. Goldacre's book is the first in a series of books looking at how pharmaceutical companies cheat in their experiments. Godfrey-Smith has one of the most beautifully written explanations of evolution, and Harari is an accessible guide to human evolution.

https://www.aqa.org.uk/resources/science/gcse/teach/command-words or https://qualifications.pearson.com/content/dam/pdf/GCSE/Science/2016/teaching-and-learning-materials/GCSE-9-1-Sciences-Command-words.pdf

The Association For Science Education. (2021). *Teaching secondary biology* (3rd ed.). Hodder Education Group.

This is a very helpful guide to teaching a range of key concepts in biology, which is strongly recommended for much more in-depth exploration than is possible in one introductory chapter such as the one provided here.

TEACHING CHEMISTRY

Ann Childs and Victoria Wong

INTRODUCTION

This chapter looks at chemistry as a school subject and focuses on two key themes.

The first theme explores the models we use to explain the properties, reactions, and uses of materials. Chemistry teachers can create what Ogborn et al. (1996) call 'a need to explain' through observations of the natural world and asking questions like:

- Why do solids have a fixed shape but liquids do not?
- Why does the reactivity of the alkali metals increase as you go down group 1 of the Periodic Table?
- Why do metals conduct electricity?

To answer each of these questions we use models to construct explanations. Many explanations in chemistry require transitions between the macroscopic or observable, the molecular and the symbolic (e.g. symbols and equations), and the requirement for these transitions is what makes chemistry so challenging. Therefore, our first theme will explore how models can act as a bridge between the macroscopic and the molecular world of chemistry and allow pupils to put on their 'molecular spectacles' (Kind, 2004) in order to explain the observations they have made of phenomena at the macroscopic level.

Our second theme is about how we use models to design and sequence explanations in the chemistry classroom. This quote from Ernest Rutherford shows how the model of the atom changed and developed as more evidence was gathered:

> It was quite the most incredible event that has ever happened to me in my life. It was almost as incredible as if you fired a 15inch shell at a piece of

DOI: 10.4324/9781003110187-7

tissue paper and it came back and hit you. On consideration, I realized that this scattering backward must be the result of a single collision, and when I made calculations I saw that it was impossible to get anything of that order of magnitude unless you took a system in which the greater part of the mass of the atom was concentrated in a minute nucleus. It was then that I had the idea of an atom with a minute massive centre, carrying a charge.

(Ernest Rutherford, cited in Pais, 1986, p. 186)

In 1899, J. J. Thomson proposed the 'plum pudding' model of the atom where the pudding was a large positive charge in which negative electrons were embedded. In 1909, Hans Geiger and Ernest Marsden, under the direction of Ernest Rutherford, performed the gold foil experiment where they fired alpha particles at a thin sheet of gold atoms. If J. J. Thomson's model was close to the actual structure of the atom, then the alpha particles should have passed straight through the gold foil, but they did not, and Rutherford's quote vividly illustrates the research team's surprise. This led to Rutherford proposing a revised model of the atom that had an incredibly small, positively charged nucleus.

In the chemistry classroom we use models to explain, and the models we use change and develop as more complex chemical concepts are explored through the course of secondary education. These changes in models demonstrate how science works and how, as observations of the natural world become more detailed, models change.

This is a very important point for chemistry teachers to emphasise because often when pupils are presented with a different model they are told that the models they have used in previous years are wrong and this can be demotivating! We want to challenge this approach, believing that we should explain to pupils that models used in previous years *are not wrong,* rather they are chosen because they helped explain the concepts *at that level* and that, as concepts become more demanding, different models are needed. As teacher educators we have come across many adults, including science graduates, who were similarly put off chemistry by being told to forget what they had learnt in previous years as it was wrong.

Chapter 11 in this book, 'Models and Modelling in Science Education', discusses the range of ways science teachers use models in the classroom and the research evidence and learning theory on which this is based. Here, the focus is on specific models you might use when teaching chemistry.

OBJECTIVES

At the end of the chapter you should be able to:

■ Understand the advantages and limitations of a range of models used to explain key concepts in chemistry.
■ Explain how and why we need to change some models for more sophisticated versions during the course of secondary chemistry education.

OVERVIEW OF THE CHEMISTRY CURRICULUM

For lower secondary pupils (ages 11-14) the chemistry curriculum would begin by considering topics such as:

- The difference between pure and impure substances
- The separation of mixtures
- The particulate nature of matter.
- Atoms, elements, and compounds, perhaps linked to the study of
- The Periodic Table
- Different types of chemical reactions including whether these reactions are endothermic or endothermic.

For upper secondary pupils (ages 14-16) key topics might include:

- The structure of the atom, linked to how elements are organised in the Periodic Table.
- Structure and bonding in elements and compounds to explain the properties of a range of elements and compounds.
- Chemical changes
- Reversible chemical reactions and chemical equilibria
- Energy changes in chemistry.

Reversible reactions, chemical equilibria, and energetics are often brought together by studying key industrial processes in the chemistry industry and explaining why the reaction conditions are chosen. For example, the Haber Process produces ammonia, an important chemical in the manufacture of fertilisers.

Key questions

When using models with pupils it is important to ask:

- What models could I use to teach this concept?
- What models am I using? Why?
- How can I help pupils to explore these models and understand their strengths and limitations?

In the rest of this chapter, we will explore some of the main models used in teaching chemistry, focusing on:

- Particle theory
- Elements, mixtures, and compounds
- Atomic structure
- The Periodic Table
- Structures and bonding.

In doing so, we will provide an overview of some important ideas in the chemistry curriculum and how we might use models to teach them.

MODELS IN THE CHEMISTRY CURRICULUM

Particle theory and models: solids, liquids, and gases

Particle models

The first model that pupils are likely to encounter in secondary chemistry is the particle model. In this model all particles are represented as identically sized inelastic spheres, sometimes known as the billiard ball model. In a solid all the particles are regularly and closely packed, touching and only able to vibrate on the spot, not move around. In a liquid the particles are able to move from place to place and are just as closely packed but are arranged in a random way. In a gas the particles are widely spaced and can move about quickly.

There are inevitably issues just about all examples of drawings of this model, including Figure 6.1. The particles of the solid are not shown as closely packed as they could be. There should be no gaps between the particles of the liquid. The particles of the gas are far too close together. Given the scale of the drawing there should only be at most one or two particles shown in the box. The gas particles could move in any direction, not just those of the arrows. The liquid particles are also moving in any direction, but no arrows are shown for them.

One way of getting around some of the problems with models on a page is to use moving models or simulations.

A useful teaching model is to use a vibrating plate attached to power pack with voltage control or a variable resistor. A see-through container is placed on the vibrating plate and filled with ball bearings or polystyrene balls. This is a great model for showing that as the energy in the system increases, by increasing the speed of the vibrations, the balls change from vibrating on the spot (solid) to moving about (showing melting to a liquid) and then leaving the liquid and flying about (boiling to a gas). This is sometimes called a kinetic theory model.

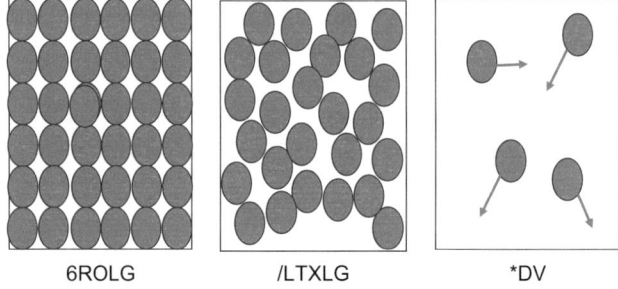

6ROLG   /LTXLG   *DV

■ **Figure 6.1** The particle model of a solid, liquid, and gas

Using the models

The important idea about the particle model, whether shown as 2D or 3D, is that it can be used to explain observations about the properties of substances. There are two important points to bear in mind in planning teaching activities where pupils use the particle model as a tool to explain phenomena. The first is that it is important to set up the *need to explain* by having pupils observe phenomena and properties of substances, particularly observations which are surprising.

For example, giving pupils the opportunity to explore sealed syringes containing a solid (perhaps sand), a liquid (water is the obvious choice), and gas (air). They will find that solids cannot be compressed. This is not surprising as they have leant on a table many times and know that it is unlikely to squash. Most are surprised, however, that water cannot be compressed at all and that the pressure of a gas prevents it from being significantly compressed as they tend to think of air as empty space. These observations of compressibility can then be explained using the particle model and, in particular, the spaces among the particles in solids, liquids, and gases.

The second important point is to support pupils in 'putting on their molecular spectacles' (Kind, 2004). In other words, providing a means of scaffolding them in moving from an observation on the macroscale (liquids are not compressible) to an explanation on the molecular level using the model. As science graduates and teachers, particularly for chemistry specialists, it is second nature to think about phenomena on the molecular level. For school pupils it is not, and they need to be guided to do so.

One way to scaffold their explanations is to have models (2D, 3D, or simulations) next to the phenomena that pupils are observing. For example, a computer simulation or the balls on a vibrating plate model next to a beaker of boiling water or a sample of sulphur being heated to melting point. Moving models are particularly powerful in supporting the move from observation to explanation, but even having a picture of the models of a solid, liquid, and gas next to the phenomena pupils are exploring is helpful. Careful questioning can be used to encourage pupils to explore and explain a range of observations using the particle model over a series of lessons. Encouraging verbal discussion through using think-pair-share can help pupils to organise their ideas and is likely to lead to better outcomes than expecting pupils to go straight to written answers.

Questions could include:

■ *Describe* what happens when you try to squeeze a solid, a liquid, and a gas in a syringe

■ Using ideas about particles, can you *explain* what you observed?

If pupils struggle, ask them to describe the model of a solid (having one they can see will support good descriptions) and then to explain why it cannot be compressed. Then describe the gas, then explain why it can be partially compressed.

Finish with the liquid, which is harder for most pupils to both describe and explain. A similar sequence of questioning can be used to support pupils in explaining a variety of different properties.

Other phenomena which can be explored and explained using the model include:

- Changes of state
- Diffusion in liquids and gases
  - Spraying air freshener in one corner of the room and asking pupils to raise their hands when they smell it
  - A potassium manganate (VII) crystal in water
  - Ammonia and hydrogen chloride gas (details in Resources)

- Mixing and separating mixtures
  - Dissolving and evaporation to retrieve the salt from a saline solution (sugar cannot be retrieved in the same way)

- Gas pressure
  - Collapsing can or bottle (details in Resources)

Concept cartoons (Naylor & Keogh, 2000), which include a description and/or picture of a phenomenon along with suggestions of a conversation some children have about it can be another useful prompt in getting pupils to discuss and explore their ideas about particles (these models are also useful in supporting pupils' use of scientific language, see Chapter 14). The Royal Society of Chemistry have an example designed to start a conversation about the condensation of water on the outside of a glass of a cold drink (see the link in the resources section).

The range of observations which can be explained using the particle model make it an extremely useful one across all three school sciences and the model continues to be used to explain a range of phenomena throughout secondary schooling up to the age of 16. There are limits to the model, however. Because all particles are treated as identical it cannot explain why different substances have different melting and boiling points. This sets up the need for a different model and theory to support explanations. The limits to this theory and the need for a new theory do *not* make the particle model 'wrong', but exploring the limitations of the model does set up the need for a more sophisticated theory to explain patterns in the properties of substances. It is worth saying again that teachers should **never, never, ever** tell pupils that what they learnt last year was wrong to set up the teaching of a new 'correct' theory.

These tasks will support you in looking at how the particle model is used in practice in your own school context.

Task 6.1 **Observe a teacher using the particle model**

Observe an experienced teacher teaching a lesson where they use the particle model to explain some observations (e.g. diffusion, gas pressure, properties of solid, liquid, and gas).

- How does the teacher support or scaffold the pupils to move from *observations* to *explanations* using the particle model?
- What questions do they use?
- How do they encourage pupils to formulate and discuss their ideas?

Task 6.2 **Planning a lesson**

Plan a lesson where you *set up the need to explain* through observations and then support pupils to move between their observations and particle model explanations. You will need to consider:

- What physical resources would you use?
- What form of the particle model would you show pupils?
- What questions would you ask? In what order?
- Will all pupils have the opportunity to discuss their ideas? How will you manage the discussion?
- How will you scaffold those questions if pupils struggle to answer them?
- How will you collect ideas and answers from pupils?

If possible, discuss your plan with an experienced science teacher.

Task 6.3 **Observe a transition to a new model**

Observe an experienced teacher teach a lesson where they transition from one model to a new one. This might be introducing the idea of elements, compounds, and atoms or the structure of the atom or moving to a more sophisticated structure of the atom model at A-level. You might like to consider:

- How do they set up the need for a new model? For example, which phenomena do they use to show why the new model is necessary? (Do they set up the need? If not, what impact does this have on pupils' ideas?)
- How do they introduce the new model to the pupils?
- How do they support pupils in using the new model to explain phenomena?
- What questions do they ask and how?

Discuss your observations with the teacher.

Atoms: elements, mixtures, compounds

The particle model, which is useful for explaining solids, liquids, and gases, does not explain elements and how they are different from compounds and mixtures and therefore a different model is required. In lower secondary school chemistry, an element is defined as a pure substance made of one type of atom only. This new model accounts for not all particles being the same. One of these particles is called an atom with, for example, the element sulphur made up of atoms of sulphur which are different from atoms in the element iron.

Compared to the particle model, there are fewer practical tasks and opportunities which allow pupils to explore the idea that different elements contain different types of atoms and the difference among elements, mixtures, and compounds. Two possibilities include iron and sulphur and/or aluminium and iodine.

Iron and sulphur

Full details of this reaction are available on the Royal Society of Chemistry website and can be found in Resources.

Pupils explore a sample of iron, sulphur, and an iron/sulphur mixture. They can put samples in water and place a covered magnet near them. The mixture is separated by both the water (iron sinks, sulphur floats) and the magnet. The key point to emphasise is that the mixture can be separated as the atoms are not joined. You could have a beaker of black beads to model the atoms of iron and yellow to model the sulphur to show that they are different.

The iron and sulphur can be heated in an ignition tube (experimental details in Resources) and the compound iron sulphide is formed. The definition of a compound is two or more elements chemically combined. The atoms cannot be separated, either by putting it in water or by the magnet. Some teachers are confused as the product of the reaction may still stick to a magnet due to the presence of unreacted iron. However, as long as the emphasis to pupils is that the atoms cannot be separated this is still a useful demonstration of the differences among elements, mixtures, and compounds. Indeed, it is one of the very few examples where it is safe for pupils to handle elements, mixtures, and compounds.

Aluminium and iodine

Details of how to carry out this experiment safely can be found in the Resources section.

This example would need to be a demonstration rather than a class practical. This spectacular reaction emits light and heat energy as the compound aluminium iodide is formed. It looks very different from its constituent elements. However, although spectacular, it has three disadvantages compared to iron and sulphur:

- The elements cannot be separated easily when mixed so it harder to demonstrate that it is a mixture.
- Although the aluminium iodide looks very different to the constituent elements there is no easy way to demonstrate that the aluminium is bonded to iodine.
- Pupils cannot handle samples of the elements, mixtures, and compounds.

Other options include magnesium and oxygen or sodium and chlorine, which must be a teacher demonstration and you should try for the first time only with a confident and experienced chemistry teacher. Details for both are in the Resources section.

Atomic structure and the Periodic Table

Atomic structure

The model of each element being made of a different type of atom is a useful one and, indeed, was the model in use by chemists towards the end of the 1800s. Dmitri Mendeleev published the first version of what would go on to become the Periodic Table in 1869 when about 56 elements were known. The simple model of each element having a different type of atom could not explain the patterns in properties of the elements and so gave impetus to the search for a more sophisticated model. There is some debate amongst chemistry teachers as to whether it is best to teach the model of the atom or the structure of the Periodic Table first. In reality, the two ideas were intertwined from the outset.

In 1914 Bohr modified Rutherford's model (discussed at the start of the chapter) by introducing the idea that the electrons were arranged in energy levels or shells. In 1932, Chadwick discovered the neutron and we now think of the atoms as having a nucleus containing protons (positively charged) and neutrons (no charge). Surrounding the nucleus are negatively charged electrons orbiting the nucleus in shells or energy levels. However, it would be wrong to think of this as the definitive model of the atom: at A-level a more sophisticated model is used and scientists are still uncovering new subatomic particles which require the model of the atom to be adjusted further.

It is worth emphasising again that the previous models of particles and atoms are not wrong, but to explain the organisation of the Periodic Table and the structures and bonding in elements and compounds at upper secondary level requires a more detailed model. This new model shows *how* the atoms of elements are different rather than just stating that they are. The new model includes the subatomic particles protons, neutrons and electrons and their arrangement. The atom has a nucleus containing protons and, usually, neutrons which is surrounded by shells or energy levels containing electrons.

Table 6.1 gives the relative masses and charges of these subatomic particles.

In the modern Periodic Table each element has an atomic number and a mass number. The atomic number is the number of protons in the nucleus and, because

■ **Table 6.1** The relative masses and charges of a proton, neutron, and electron

Subatomic particle	Where it is found	Relative mass	Relative charge
Proton	Nucleus	1	+1
Neutron	Nucleus	1	0
Electron	Shells	Very small	−1

■ **Table 6.2** The maximum number of electrons each shell can hold for the first 20 elements in the Periodic Table

Shell or energy level	Maximum number of electrons in this shell
First	2
Second	8
Third	8

atoms have no charge, also the number of electrons. For example, sodium has an atomic number of 11 and so has 11 protons in its nucleus and 11 electrons.

The mass number is the number of protons plus the number of neutrons. As the relative mass of electrons is tiny, they do not significantly contribute to the mass of atoms. Sodium has a mass number of 23 so with 11 protons it must have 12 neutrons to make up a mass number of 23.

The electrons in atoms occupy energy levels or shells which surround the nucleus; different shells can hold different maximum numbers of electrons. Table 6.2 shows the maximum number of electrons each shell can hold for the first 20 elements in the Period Table.

The electrons in an atom first occupy the lowest available energy level, which is nearest to the nucleus. When the first shell is full with two electrons, subsequent electrons will occupy the next shell and when that is full (eight electrons) the third shell will then be occupied. Figure 6.2 shows structures of the atom for sodium and fluorine.

The model for sodium shows it has 11 protons and 12 neutrons in the nucleus. It also shows that it has 11 electrons arranged with 2 electrons in the first shell, 8 in the second shell, and 1 in the third shell. This is written as 2,8,1 and is called the electron configuration of sodium. Electron configurations written as numbers are much easier for pupils to understand once they have gained experience of drawing electronic structure diagrams or doing the task suggested later.

Using the models of the atoms of sodium and fluorine we can now explain why atoms of elements are different. Sodium atoms always have 11 protons and 11 electrons and fluorine atoms always have 9 protons and 9 electrons.

Note that drawings of diagrams of the structure of atoms can vary considerably between different textbooks and websites. For example, electrons in some of these models will be represented as dots (as in figure 6.2) or they can be

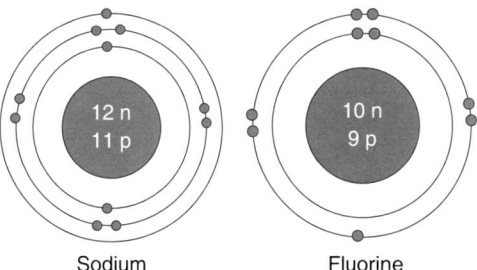

Sodium   Fluorine

■ **Figure 6.2** Structures of the atom for sodium and fluorine

drawn as crosses. The colours they are shown in can also vary. These differences can be confusing for pupils and so when choosing how you draw the structure of the atom, do look at how textbooks, websites, the exam papers, or even other teachers in the science department of the school draw them and discuss this with your pupils, so they are aware of the multiple ways we can draw the structure of the atom in chemistry.

This model can help to establish why atoms form charged ions if they gain or lose electrons. Atoms have equal numbers of electrons and protons and so are uncharged. When they gain or lose electrons they become charged as the numbers of positive and negative charges is no longer the same. It can also be used to describe isotopes as the numbers of neutrons in the atom can vary. As long as the number of protons stays the same, the atom is the same element. This is different to the model of the atom used for elements, mixtures, and compounds as it accounts for some atoms being the same element but with different mass numbers.

This model does not explain the relative sizes of the subatomic particles very well. The nucleus is tiny compared to the size of the rest of the atom and holds almost all of the mass. Electrons are far smaller than protons and neutrons but usually all three subatomic particles are represented in books by dots or crosses of approximately the same size. A different model to get a better sense of the size of the nucleus in the atom is visualised by considering a football in the centre of a large sports stadium such as Wembley stadium. The stadium represents the atoms as a whole and the football in the centre of the stadium's pitch represents the relative size of the nucleus. Again it would be important to emphasise to pupils that the model we use for the structure of the atom is not wrong, it is just that to explain the relative size of the nucleus needs a different model.

It can help pupils to understand and describe the model if they can interact with it. Models can be as simple as an outline structure drawn on a piece of paper and different coloured counters or paper circles to represent subatomic particles. It is also possible to buy plastic versions of this model (see Resources). Pupils are asked to 'make' models of different atoms from the atomic and mass numbers.

This physical representation of the model of the atom has the following advantages:

■ It allows for a more hands-on activity, supporting pupils to appreciate where the different particles go in the atom and how the electrons are structured.
■ It can be useful for establishing why removing an electron from a sodium atom forms a one positive, Na^+, ion, an idea many pupils find difficult. The model protons can be lined up and paired with the model electrons to show there is one more proton (positive charge) than electron (negative charge) and the overall charge on the ion is 1^+.
■ It can be useful for making models of the isotopes of elements. For example, one group could make a model of the isotope of chlorine which has 18 neutrons in its nucleus and another group the one with 20 neutrons and compare them.

Task 6.4 Critically evaluating modelling the structure of the atom

■ Do you agree with the strengths of the modelling activity described above?
■ Does it have other strengths?
■ What disadvantages would this activity have? Consider for example the classroom management issues and how you might need to plan to deal with them. How does this modelling activity explain why elements are different?
■ What does it not explain? Or what are the limitations?
■ What possible misconceptions might it introduce?
■ How might you improve or adapt this activity to address the challenges you have identified?
■ Ask science teachers in your school about the activities they use to introduce the structure of the atom.

Atomic structure and the Periodic Table

The Periodic Table was developed by scientists in the late 1800s. The work of Dmitri Mendeleev is significant because he used the emerging Periodic Table to predict the properties of undiscovered elements, for example, an element which he named as eka-aluminium. The element was eventually discovered by Paul de Boisbaudran in Paris in 1875. He named the element gallium after the Latin name for France. Table 6.3 shows just how remarkably close Mendeleev's prediction was to gallium's actual properties, which illustrates how powerful these early attempts at producing a Periodic Table were in being able to predict the properties of as yet undiscovered elements.

Although these early chemists could describe many of the physical and chemical properties of elements and then group them, they could not explain why elements in the same group, for example, Group 1, the Alkali Metals or Group 7, the Halogens, had similar physical and chemical properties. Their achievements in working out the structure of the Periodic Table is therefore all the more remarkable.

■ **Table 6.3** Mendeleev's prediction compared with Gallium's properties

Property	Property Eka-aluminium (Mendeleev's prediction)	Gallium
Atomic mass	68	69.72
Density (g/cm3)	6.0	5.9
Melting point (0C)	Low	29.78
Formula of oxide	Ea_2O_3	Ga_2O_3
Formula of chloride	Ea_2O_3	Ga_2Cl_6

■ **Table 6.4** Relating the models of sodium and fluorine to their group, period, and atomic number

Feature of the Periodic Table	Relationship to the structure of the atom of sodium	Relationship to the structure of the atom of fluorine
Group number	Sodium has one electron in its outer shell and is in Group 1 of the Periodic table.	Fluorine has seven electrons in its outer shell and is in Group 7 of the Periodic table.
Period	Sodium has electrons in three shells with an electronic configuration, 2,8,1. This means it is found in the third period of the Periodic Table	Fluorine has electrons in two shells with an electronic configuration, 2,7. This means it is found in the second period of the Periodic Table
Atomic number	Sodium has 11 protons in its nucleus and atomic number is usually defined as the number of protons in the nucleus. Atomic number of sodium = 11 Sodium is the eleventh element in the Periodic Table and elements in the Periodic Table are arranged in order of their atomic numbers from left to right	Fluorine has nine protons in its nucleus and atomic number is usually defined as the number of protons in the nucleus Atomic number of fluorine = 9

The structure could only be explained once the structure of the atom was sufficiently understood.

The model of a nucleus containing protons and neutrons surrounded by electrons arranged in shells or energy level allows the explanation of:

1 Group number
2 Period number
3 Atomic number

in the Periodic Table. Let's take the elements sodium and fluorine as examples again (see Figure 6.2).

Table 6.4 below shows how the models of sodium and fluorine can give information about which group and period of the Periodic Table they are in, and their atomic numbers.

The model can also explain why, for example, the elements in a group have similar physical and chemical properties: they have the same number of electrons in their outer shell. For example, all elements in Group 1, the Alkali Metals, have one electron in their outer shell and all the elements in Group 7, the Halogens, have seven electrons in their outer shell. It is the electrons in the outer shell of the atoms of elements that are involved in chemical reactions and in chemical bonding. Therefore, elements with the same number of electrons in their outer shell in groups 1, 2, 7, and 8 have similar physical and chemical properties. Demonstrating the similarities of the properties is most straightforward with the alkali metals.

The reactions of the alkali metals with water

This is a classic demonstration, and full details, including the equations for the reaction and risk assessments can be found on the RSC website (see link in resource list).

This demonstration is important to establish that the chemical reaction of the alkali metals with water produces the same products: hydrogen gas and the metal hydroxide. The reaction is the same for each alkali metal because they all have one electron in their outermost shell. It is the outer shell electrons which are rearranged during chemical reactions.

However, this demonstration also shows that, as the group is descended, from lithium to potassium the reactivity of the alkali metals increases. This increase in reactivity can be further consolidated by showing a video (such as the one from the Open University suggested in the resource list) which shows lithium, sodium, and potassium being added to water thus consolidating the demonstration. It goes on to show rubidium and caesium reacting with water and the video could be paused before rubidium and caesium to ask pupils to predict what they think will happen.

Demonstrating this pattern in reactivity can be used to set up a need to explain the pattern. Pupils can be asked to draw atoms of lithium, sodium, and potassium. These models show that the outer electron from lithium to sodium to potassium gets further and further from the positive nucleus so there is less attraction between the outer negative electron and the positive nucleus and, therefore, the outer electron is lost more easily and the reaction occurs more vigorously.

Task 6.5 **Evaluating activities**

Questions
- What are the advantages and disadvantages of doing the demonstration – what does it explain? What are its limitations?
- What are the advantages and disadvantages of showing a video? What does it explain? What are its limitations?
- If you were explaining the reactions of the alkali metals would you use the demonstration, a video, or both? Why? What would you need to consider when making your decision?

Task 6.6 **Planning a sequence of questioning**

Watch the video of the reactions of lithium, sodium, and potassium.

■ What would you *tell* pupils as you carried out the demonstration?
■ What *questions* would you ask to support pupils in using the model to explain the observations?
■ How would you ask the questions and how would you gather answers? For example, would you use written questions, think-pair-share, mini white boards, targeted questioning, hands up questioning? What are the advantages and disadvantages of each? Ensure you can justify your choice.

Structure, bonding, and the properties of matter

The model of the atom with its protons, neutrons, and electrons is essential in explaining the formation of the different structures of elements and compounds summarised in Figure 6.3.

The structure of metallic elements is often described as a 'giant metallic structure of ions'. The outer electron(s) of the metal atoms are delocalised to form a 'sea' of electrons. The atoms have now lost electrons and are therefore positive ions.

Figure 6.4 shows two possible models for the metallic structure which could be used to explain the physical properties of metals, for example that they conduct electricity, have high melting points, and are malleable (can be shaped by being hit).

■ Which model would you use? Why?
■ What are the advantages of each model?
■ What possible limitations does each have?
■ What misconceptions might each promote?

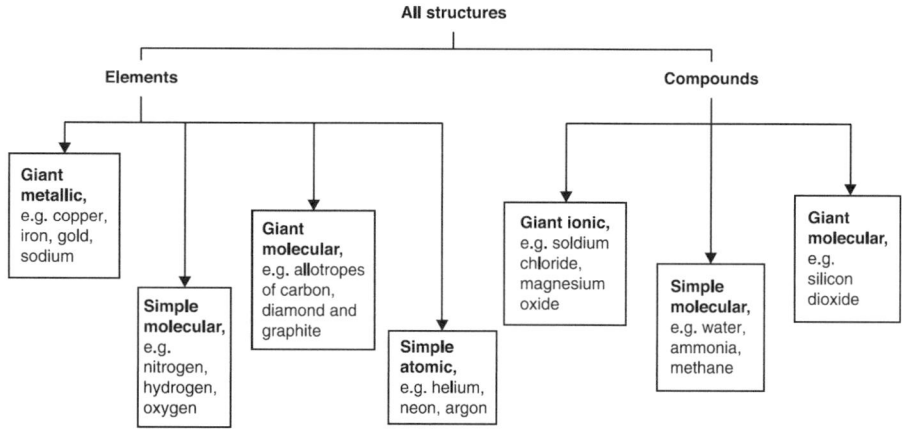

■ **Figure 6.3** The different structures of elements and compounds

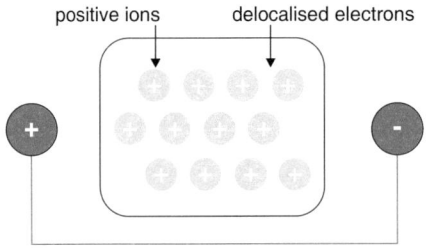
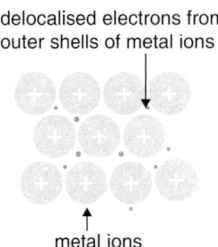

■ **Figure 6.4** Models of giant metallic structure

Neither of these models is fully correct and neither is fully wrong but it is import-ant, as argued earlier, for chemistry teachers to be clear about which model they are choosing to explain observations, what advantages the model offers for the explanation, what its limitations are, and what possible misconceptions it might promote. The science curriculum for 14–16-year-olds in England also requires pupils to be able to identify advantages and limitations of models of structure and bonding.

SUMMARY AND KEY POINTS

In summary, models are useful to explain key chemical ideas and when pupils are presented with a different model they are often told that the models they have used in previous years are wrong! We have challenged this approach and have taken the view that we should explain to pupils that models used in previ-ous years *are not wrong*, rather they are chosen because they helped explain the concepts *at that level* and that, as concepts become more demanding, differ-ent models are needed. Indeed, the story of the development of ideas about the atom demonstrates that this is how science works: more observations and data lead to more sophisticated models.

In addition, we have emphasised that before using a model with pupils it is important to assess the strengths of the model, its weaknesses, any possible misconceptions it might promote, and then make an informed decision about whether to use it. This may make us abandon the model for another choice but, if it does not, we are at least aware of some of the limitations the model has and how we may take account of this when we use it with pupils.

Such teaching will allow us to present chemistry as a dynamic and evolving discipline rather than a subject which requires the learning of content teachers know is wrong.

Check which requirements for your initial teacher education you have addressed through this chapter.

FURTHER RESOURCES

Collapsing bottle experiment. https://spark.iop.org/collapsing-bottle

Diffusion of ammonia and hydrogen chloride gases. Retrieved July 25, 2023, from https://edu.rsc.org/experiments/diffusion-of-gases-ammonia-and-hydrogen-chloride/682.article

Kinetic theory model. Retrieved July 25, 2023, from https://www.philipharris.co.uk/product/chemistry/materials-and-their-properties/states-of-matter/kinetic-theory-model/b8h25365

Plastic atom models (Bright atom). Retrieved July 25, 2023, from https://www.philipharris.co.uk/product/chemistry/materials-and-their-properties/structure-and-bonding/bright-atom/b8r07169

Reacting aluminium and iodine. Retrieved July 25, 2023, from https://edu.rsc.org/experiments/reacting-aluminium-and-iodine/715.article

Reacting magnesium and oxygen. Retrieved July 25, 2023, from https://edu.rsc.org/experiments/the-change-in-mass-when-magnesium-burns/718.article

Reacting sodium and chlorine. Retrieved July 25, 2023, from https://edu.rsc.org/exhibition-chemistry/the-reaction-between-sodium-and-chlorine/4015463.article

Reaction of a large piece of sodium with water. Retrieved July 25, 2023, from https://edu.rsc.org/resources/ri-christmas-lectures-2012-the-alkali-metals/1124.article

Reactions of the alkali metals. Retrieved July 25, 2023, from https://www.youtube.com/watch?v=6ZY6d6jrq-0

RSC concept cartoons. Retrieved July 25, 2023, from https://edu.rsc.org/resources/condensation/1869.article

The following examples are helpful practical details and videos referred to in the main text. The RSC has many other examples of activities, videos, and resources for use in teaching chemistry in school.

The reaction of iron and sulphur, practical details. Retrieved July 25, 2023, from https://edu.rsc.org/experiments/iron-and-sulfur-reaction/713.article

The RSC have produced a useful interactive Periodic Table, which can be engaging for pupils to use and explore. https://www.rsc.org/periodic-table

Video of Group 8 or 0, the noble gases. https://edu.rsc.org/resources/ri-christmas-lectures-2012-the-noble-gases/1127.article

TEACHING PHYSICS

Jill Noakes

INTRODUCTION TO THE CHAPTER

There is a well-known quote about physics, often mistakenly attributed to either Albert Einstein or Richard Feynman, which says if you can't explain something to your grandmother, you don't really understand it. The exact origins of this statement are ambiguous (and quantum physicists might not agree!) but the sentiment behind it strikes a chord with many who are faced with explaining electric current, Newton's Third Law, or electromagnetism to a classroom full of pupils for the first time.

Learning to teach physics will challenge your very notion of what it means to understand something. You may *think* you understand it, but as your explanation unfolds you begin to notice gaps and incongruences, or worse, these are highlighted for all to see by some innocent but painfully astute question. Alternatively, you may not feel that you understand it that well in the first place, so you stick to the textbook, muddle through as best you can, and store up their questions to ponder after the lesson. There is of course a third possibility which few consider, that you may in fact understand it *too* well, and have entirely forgotten what it was like not to know what an electron is.

Whatever your starting point, communicating your understanding and enthusiasm clearly through dialogue, imagery and practical experiences, in a way that builds securely on what the learner already knows, is a skill that takes time to develop. None of these observations is unique to physics, but in many ways physics brings out the problems of being a new teacher in sharper relief than almost any other subject. This chapter will introduce you to some of the key topic areas in the physics curriculum which have been chosen to illustrate some particular conceptual and pedagogical challenges. It also offers a chance to reflect on what learners experience in the physics classroom and how we as teachers can provide a supportive and inclusive environment.

DOI: 10.4324/9781003110187-8

OBJECTIVES

At the end of this chapter you should be able to:

■ Reflect on the paths to effective physics teaching
■ Consider how a selection of key concepts in physics is addressed through the curriculum
■ Discuss some of the socio-cultural challenges of teaching and learning physics and possible directions for addressing them.

BECOMING A TEACHER OF PHYSICS

The first step on any path to becoming an effective physics teacher is self-reflection. Whether you are conscious of it or not, the sort of physics educator you will become will be at least partially informed by your past experiences as a learner as well as your current ones as a student teacher. Millar (1988) found that many biology and chemistry student teachers had consciously 'opted out' of physics at an early stage in their career, and this led them to anticipate difficulties with teaching a wide range of physics topics. Although Millar's study was some years ago, many student teachers still report similar experiences and anxieties, so it seems little has changed. Even with full command of the subject matter, it is common for student teachers to experience a range of subtle challenges and transformations as they progress, as Findlay and Bryce (2012) put it, from thinking about teaching *physics* to teaching *children* (about physics). Before you start teaching it is useful to reflect on your own experiences of learning physics. Task 7.1 asks you to take a few moments to look back and reflect on how your experiences as a learner might inform your practice as a teacher.

Task 7.1 **Looking back, thinking forward**

Either

If you have had really positive experiences of learning physics, try to reflect on what the teacher(s) did to keep you engaged and interested. What was the pace like? Did they use clear language and imagery? Were your classmates equally engaged, or were any of them left behind? Pick out the best aspects of their practice but also try to think realistically about whether their approach could have been improved in any way, particularly for others who might not have shared your enthusiasm for the subject.

　OR

　If you feel generally quite negative about your experiences of learning physics, try to pick out what it was in particular that made it difficult or unenjoyable. What do you wish your teacher(s) had done differently? Is there anything you wish you could have told them while they were teaching? Now consider how you can turn these experiences into positive action to make your own pupils' learning journey more accessible.

　Discuss your experiences with other student teachers. It doesn't matter if they are not learning to teach science, in fact teachers from other subjects might have some different perspectives that are worth listening to.

PROGRESSION IN THE PHYSICS CURRICULUM

The physics curriculum varies between countries, and how schools choose to implement it can also vary considerably. However, the 'big ideas' that underpin physics are widely shared, and many of the difficulties that learners encounter transcend local or cultural boundaries (Ozdemir, 2017; Butts, 1985). Although physics is frequently taught as a series of unconnected facts and theories, gaining a sense of the over-arching grand narratives and unifying concepts is essential to instilling the awe and wonder that we want our pupils to share. In Chapter 8 Paul Davies explores the challenging task of planning for progression in science generally, and physics is a particularly hierarchical subject. While there may be different routes through any given topic, the connections both within the topic and to other topics must always be made clear. For example, radioactivity appears in the later stages of most physics curricula because it requires a firm understanding of atomic structure from chemistry as well as electromagnetic waves, the concept of electric charge, ionisation, and general ideas about energy. Planning a teaching sequence in physics should always take account of what learners already know and where they will be going next.

Rogers (2018) lists five 'big ideas' in physics as energy, electricity, forces, particles, and the universe, and this loosely matches up to the Big Ideas identified by the ASE (Harlen, 2010). While other texts such as Sang (2011) offer detailed teaching sequences for each of these themes, here I will discuss just three of them in order to highlight some of the issues faced by early career physics educators. In doing so this chapter also provides physics-specific examples of some of the more general ideas discussed elsewhere in this book. First, the discussion of energy highlights the importance of language and how it shapes our imagery and understanding of physics (see also Chapter 14). The teaching of forces provides a chance to reflect on the careful construction of a very specific picture of reality (Chapter 10), and the section on electricity emphasises the centrality of models in our understanding of the invisible and abstract (Chapter 11).

Energy

Energy makes frequent appearances throughout the secondary curriculum, and many topics such as heat, light, and electricity make use of the concept of energy, even though they often seem to pupils to be entirely unrelated to one another. I have chosen to start with energy for two reasons. First, it encompasses absolutely *everything*. There isn't a single phenomenon in the universe for which questions about energy may not be asked. Second, it is the slipperiest of physics concepts, so slippery in fact that it doesn't exist (at least not in any material sense). However, as non-existent things go, it is incredibly useful. The concept of energy gives rise to measurements and calculations that help us to make the most of fuel resources, to predict temperature changes, to power our gadgets, and so much more.

In the late eighteenth and early nineteenth centuries it was believed that there was a substance called 'caloric' which flowed from hot objects to cooler objects and made things happen. We now know that there is no physical substance, no invisible fluid or observable 'stuff' that literally moves from one place to another. However, the mathematical tools and ways of understanding the physical world that stemmed from this idea are still incredibly effective for calculating, predicting, and making sense of pretty much any process you can think of. For that reason, it is still helpful and in fact necessary to speak about energy as if it really is a 'something'.

However, a debate that has been taking place for decades and that has only very recently reached a new consensus among exam-setters and textbook writers (in England at least) is whether energy is one type of 'something' or many. To cut a long story short, it is now considered more conceptually sound to think of energy as a single type of 'something' that is transferred between different kinds of stores (thermal stores, gravitational stores, etc.) rather than talking about different 'forms' of energy that somehow transform into one another when something happens.

Example – a kettle on a gas camping stove
Forms of energy: Chemical energy is transformed into thermal energy.
Energy stores and transfers: Some of the energy from the chemical store of the fuel is transferred via heating to the thermal store of the water.

At first glance this may look like a superficial change of wording, but we are not simply swapping one set of labels for another. In Chapter 12 Judith Hillier explores the importance of language in science and this is a good example. In the first account it is difficult to see why there is any need to label the energy; one might as well say that the red energy turns into blue energy before ending up as purple energy (Millar, 2014). Note that in the second explanation there are two parts to describing each store: 'the _____ store of the _____' so we are identifying not only the type of store but also being specific about which particular object(s) to focus our observations on. This is particularly important with regards to thermal stores, as 'thermal energy' fails to distinguish between energy that is transferred to the water in the kettle and energy that is transferred to the surroundings.

The second description also evokes a mental image of one store emptying while one or more stores elsewhere fill up by the same amount, thus making the idea of conservation of energy seem quite intuitive. Selecting a start and end point is essential so that it is possible, at least in theory, to make measurements and calculations regarding how much energy is stored in the fuel before the stove is lit and how much remains at a particular point in time later. Similarly, one can calculate the increase in the thermal store of the water over the same specific time by measuring an increase in temperature.

The fundamental point of the stores and transfers approach is to focus attention on *quantities*, for without quantification the concept of energy is meaningless. Applying labels to different stores is not an end in itself but a way of identifying which tools to use to calculate *how much* energy is present at different points in a given process. Moreover, energy stores afford us a snapshot view of the distribution of energy at a given time, although this is only really useful if one considers a 'before' and 'after' snapshot in order to find out what has changed.

These are the energy stores that are discussed in most (recent) textbooks, although some of these are not really considered much until A Level:

- Kinetic stores
- Gravitational stores
- Chemical stores
- Elastic stores
- Thermal (internal) stores
- Nuclear stores
- Electrostatic stores (e.g. a capacitor)
- Magnetic stores (e.g. the energy stored when two repelling poles are pushed together).

Three things that are conspicuously absent from this list are electricity ('electrical energy'), light and, sound, which were all traditionally considered to be forms of energy. Under the new consensus approach, these are not ways in which energy is *stored* but rather *pathways* by which energy is transferred from one store to another. Identifying and understanding these active processes such as electric current (work done by charged particles moving in an electric field or 'electrical working'), waves (electromagnetic, sound, water waves, etc.), mechanical working (by means of physical forces) and heating (conduction, convection, and radiation)[1] enables us to ask and answer questions about *efficiency*. Where in the process is energy being diverted to stores that are not useful to us? How can this be reduced in order to ensure that as much energy as possible is transferred to the stores that *are* useful to us? The transfer pathways also lend themselves to measures of the *rate of transfer* of energy, otherwise known as *power*. It is not meaningful to ask how much energy exists in the form of light at a given moment, but one can ask important questions about how much energy is being transferred per second by light waves emanating from a given source.

The science of energy is ultimately a mathematical one that is meaningless without measurements and calculations, but in order to know which measurements and which equations to use, we must begin with storytelling. Task 7.2 will give you an opportunity to practice telling stories about energy and using the appropriate language so that you can more confidently teach others to identify stores and transfer mechanisms.

Task 7.2 **Telling the energy story**

Think of a specific example from an everyday context that you could show to pupils or get them to do themselves (e.g. something being heated, something being moved, something battery operated) and make sure you can tell the story of what energy changes take place using the language of stores and transfers. Give an account of a) which store(s) is/are emptying, b) which store(s) is/are filling, and c) in loose terms (don't get hung up on too much detail) what kind of transfer process(es) is/are involved. You must choose an appropriate start and end point otherwise your story could go on forever! Now consider how you would structure a lesson to help pupils construct their own energy stories using the language of stores and transfers.

The final important point to note about energy is that it does not *cause* things to happen. If we want to understand why things happen we must look at forces and other dynamic processes in the universe. Energy calculations simply allow us to determine whether or not something is *possible*. In that sense it is a bit like money: without it you cannot make things happen, but having lots of it does not determine what happens next. Unlike money, energy is always conserved, and this of course is one of the ultimate unifying principles of science.

Forces

'Forces' is another topic area that appears in fragmented ways throughout the secondary curriculum. For instance, a unit about 'space' will often feature lessons on mass and weight, rocket propulsion, and so on, and it is important that pupils are taught to recognise that this is the same set of principles as the ones they encounter in their units on 'Forces and Motion'. Much of it seems like fairly safe territory; Newton and his contemporaries had it pretty much figured out and very little has changed (bar the discovery of a few new types of force), so what could go wrong? The trouble with teaching and learning about forces is that there is very little to remember – Newton's Laws of Motion along with a handful of equations and definitions – but an almost infinite variety of different scenarios, each requiring the following thought process to be followed:

1 What are the forces acting on the object of interest?
2 Which direction do they act in and how big are they?
3 What is the net result of adding these forces together?
4 What effect, if any, does this resultant force have on the object's motion?

Thus memorising equations and laws is only the beginning, the acquisition of a set of tools. Using these tools to answer questions and analyse situations that seem to defy common sense is a skill that takes time, which is why forces are revisited at every stage of the curriculum. The challenge is compounded by the fact that understanding forces requires the learner to look at the physical world in a special way, as if through a special pair of 'force goggles' – much like the 'molecular spectacles' idea mentioned in Chapter 6: Chemistry. Force goggles focus on one object at a time, transform irregularly shaped objects to much simpler ones, and completely remove details that would make little difference to the analysis (a judgement that is not always obvious).

Common misconceptions around forces and motion are well researched and documented (e.g. Driver et al., 2005; Ozdemir, 2017) and there is also evidence that misconceptions arise spontaneously in response to new scenarios rather than always deriving from a pre-existing mental framework that the learner brings with them (Rowlands et al., 2007). This makes it particularly tricky to guide pupils towards a coherent theoretical schema that can be reliably applied in new and unfamiliar situations. Hence, it is important not only to focus on the basics of constructing force diagrams – calculating resultant forces and accelerations – but to frequently model how to analyse and make sense of surprising and counter-intuitive scenarios in a qualitative or semi-quantitative way. Task 7.3 offers a selection of some scenarios for you to test your force analysis skills, with an emphasis on talking them through with a colleague.

Task 7.3 **Fluency with forces**

Pick at least one of the following questions and try to develop a full and clear explanation suitable for a lower secondary audience. You might look online for a suitable video or website, find or draw appropriate diagrams or images, and think carefully about what key words and concepts pupils would need to be familiar with in order to make sense of it.

Why do a hammer and a feather fall at the same rate on the moon but not on Earth? (Hint, you first need to convince learners that this is in fact the case, several videos are available online to demonstrate.)

Why do astronauts train for space missions under water? (Think about the forces, not just the lack of air.)

Why can we ignore air resistance in some situations but not others?

When you have developed your explanation, share it with your school mentor or subject tutor to see if it needs to be refined, simplified, or altered. Do not be embarrassed if it turns out your explanation is incorrect or incomplete; now is the best time to fill in the gaps or misconceptions in your knowledge, not when you are facing a class.

For further reading you may wish to refer to Watts (2014) for a general discussion of what it means to really 'explain' something in science teaching.

Electricity

In primary school, pupils learn to build and manipulate simple electric circuits without the need for any theoretical underpinning. They also have years of everyday experience of using electrical devices and hearing electricity spoken about in non-scientific contexts. By the time formal concepts such as current, potential difference (voltage), and resistance are introduced at secondary level, each and every pupil will already have some kind of mental picture of what is happening inside electrical wires, whether they are conscious of it or not (Osborne, 1983; Driver et al., 2005; Chapter 10 of this book). They may not be able to articulate it, but they will likely use it to infer certain things about electric circuits which feel intuitively correct but are scientifically misleading. It is entirely possible to ignore these assumptions and teach pupils to memorise definitions of current, potential difference, and facts about what happens to them in series and parallel circuits, without ever equipping them with an even broadly scientific model of what is really happening. However, the greater challenge is to accompany this with a gradual shift away from unhelpful conceptual images towards those that allow them to make sense of it all in more meaningful ways.

There are many animations, images, and pupil models that offer visual representations of what happens inside electric circuits. Some are excellent while others can encourage decidedly unhelpful misconceptions. However, among the many models that are available, there is one that is both powerful and staggeringly simple. It requires nothing more than a loop of rope and is easier to demonstrate than explain, but if there is no one on hand who can demonstrate it to you, an introductory description can be found by searching for 'rope loop electric circuit model' on IOPSpark (see Further Resources)

Task 7.4 **Electrical misconceptions**

Take some time to explore the Institute of Physics (IOP) Spark website. For this task, focus on finding out about common misconceptions around electricity and evaluating some of the diagnostic questions that are provided (look for the Misconceptions section and select the Electricity and Magnetism domain). Note that the relevant research literature is referenced below each misconception. Next, follow some of the links to resources that are suggested to address particular misconceptions and make a note of anything you find that is particularly useful.

If you get the opportunity during your teaching practice, you could consider using one or more of the diagnostic questions to examine the prevalence of a particular misconception among your pupils and then assess their understanding again after a well-planned teaching intervention based on the suggested resources and ideas (with input from other sources as appropriate). This could be an informal exercise built in to your day-to-day practice or part of a small research project.

Writing notes on this task will help you to address a number of key areas in your training.

General issues with practical work in science are discussed in Chapter 13, but it is worth noting here that electricity is one of the main topics in which practical work can sometimes obscure and even hinder conceptual understanding rather than reinforcing it. Small uncertainties in meter readings can be interpreted by pupils as 'proof' that their misconceptions are correct (for instance that current really is used up) and the logistical challenges of building a functioning circuit can easily occupy the entirety of pupils' attention such that the point of the practical may be completely lost.

Research comparing real 'hands-on' practical work with virtual experiments in electricity has suggested that virtual models can be a more effective way to embed conceptual understanding of current flow and related ideas as they allow the learner to 'see' inside the circuit without the distraction of tangled wires and faulty bulbs (Zacharia & de Jong, 2014). There are some really excellent interactive resources available (e.g. PhET's Circuit Construction Kit, see Further Resources) and when it comes to teaching concepts these are an invaluable tool. Of course we also want to teach pupils to handle electrical apparatus safely, to construct circuits confidently, and take measurements, but it is probably wise to be clear that these skills are the focus of most practical work in the early secondary phase and not to expect the conceptual understanding to arise spontaneously or to be in any sense 'obvious' from the results of a practical.

BROADER ISSUES IN PHYSICS EDUCATION

There is far more to the physics curriculum than energy, forces, and electricity, but the areas discussed here have been carefully chosen to illustrate some broader principles. The discussion of energy highlights the importance of language and how it shapes our imagery and understanding of physics. In forces, everyday experience meets the scientific worldview in often spectacularly confusing ways, which emphasises the need to unpick misconceptions and even provoke new ones before attempting to tackle them in a careful and structured way. Finally, electricity exemplifies both the power of a good model and the challenges of trying to build practical skills and conceptual understanding of an invisible and highly abstract concept at the same time (See Chapters 14, 10 and 11 for more discussion on language, constructivism, and modelling in science teaching).

However, there are some issues related to teaching physics that are social and cultural in nature. These are related to broader issues in science education and not necessarily unique to physics, but it is fair to acknowledge that, culturally speaking, physics has a bit of an image problem. After interviewing 132 15-16 year olds and their parents in the UK, Francis et al. (2016) found that physics in schools is widely regarded as a 'hard' subject and, furthermore, that it is frequently represented as a subject 'for men'. In some parts of the world, for example in England, it is the least popular science subject in post-compulsory education and has the worst gender balance by far (Ofsted, 2015). Although wider issues regarding teaching for social justice are explored in-depth in Chapter 18, there is a body of research and a variety of ongoing initiatives that relate specifically to participation in post-compulsory physics.

The issues are numerous and complex, and Murphy and Whitelegg (2016) offer a broad review of the evidence as it specifically relates to physics. While some of the barriers are deeply rooted in societal and parental expectations (Archer et al., 2012) and others have been shown to be heavily influenced by whole school factors (IOP, 2012, 2013), individual teachers can have a significant impact, either for better or worse. A sample of teachers in Sweden were found to grade work more harshly if they were told that the author of the work was a girl (Hofer, 2015). Furthermore, inexperienced teachers were more likely to demonstrate bias than more experienced colleagues. In more general terms, Mujtaba and Reiss (2013) found that girls often have very different experiences in school physics lessons than their male peers and are less likely to be encouraged by their teacher to continue studying physics.

These findings are illustrative rather than comprehensive and may not be generalisable (it is certainly not the case that all teachers display gender bias), but they raise important questions and suggest some small changes that could begin to make a difference in your own classroom. For example, thinking about how you employ questioning strategies to avoid certain pupils dominating the discussion or how roles are allocated within groups so that everyone has an equal chance to handle apparatus. The sort of practices that might arise from an awareness of these issues, such as the avoidance of stereotypes, positive encouragement, varied approaches to teaching and assessment, and so on are basically just good teaching and should in fact benefit everyone. It is therefore not a case of disadvantaging boys to benefit girls but rather trying to remove some of the invisible barriers that prevent many pupils (male and female) from enjoying physics and aspiring higher.

SUMMARY AND KEY POINTS

This chapter has offered a brief overview of some of the issues involved in learning to teach physics.

- Whatever your subject background, it is important to reflect on your strengths and weaknesses with regards to physics, take suitable measures to overcome anxieties, and/or identify ideas you take for granted
- Energy as a unifying concept is important throughout the curriculum, but ensuring a clear, consistent, and logically coherent approach to the language and imagery employed is essential
- Forces are another fundamental way of understanding the physical world, but an area of the curriculum that is rife with misconceptions and often at odds with 'common sense'
- Electricity is both conceptually and practically challenging to teach. Think carefully about the models and analogies you use, and plan your practical work with clear emphasis on which particular knowledge or skills you want pupils to develop

■ Despite some progress, physics education still faces some cultural challenges. As an individual teacher you cannot fight these alone, but there are small changes you can make to your own practice that will benefit everyone. There is also research available that can help to inform dialogue and development on a departmental or whole-school level.

Check which requirements for your initial teacher education you have addressed through this chapter.

NOTE

1 Thermal radiation, otherwise known as infrared, could be listed under Waves or Heating. These categories are not set in stone and different curriculum specifications may divide them up and label them in different ways. It is not the classification system that matters but the understanding of the processes themselves.

FURTHER RESOURCES

DeWinter, J., & Hardman, M. (2021). *Teaching secondary physics* (3rd ed.). Hodder Education. This book provides some strategies and suggestions for teaching each of the main topic areas of physics. Each chapter includes an indication of what pupils might already know, suggested teaching sequences, likely misconceptions and hints about practical work, as well as further links to other relevant resources.

IOP Spark https://spark.iop.org/. A treasure trove of pedagogical content knowledge, including common misconceptions and how to address them (all linked to research references), detailed guidance for practical work, diagnostic questions, and much more. You will also find links to TalkPhysics, the online community for physics educators, and a range of other resources and events offered by the Institute of Physics.

Isaac Physics https://isaacphysics.org/. For GCSE and A Level this is a fantastic resource for developing your own subject knowledge and setting work for your pupils. Based around the principle of mastery, it comprises sets of carefully structured questions that are answered and marked online, supported by a wealth of subject guidance, hints, and problem-solving resources. Isaac Physics offers online and face-to-face support and CPD for both student teacher and in-service teachers.

PhET Interactive Simulations https://phet.colorado.edu/. This site is packed with engaging and interactive simulations. A few favourites include the Circuit Construction Kit, the Energy Skate Park, and Forces and Motion: Basics, but exploring the rest of the site is thoroughly recommended.

Rogers, B. (2018). The big ideas in physics and how to teach them. In *Teaching physics* (pp. 11–18). Routledge.

This book is an example of how learning theories and cognitive science can inform (one approach to) the teaching of specific physics topics. It offers over-arching historical narratives to bring together the five big themes (Electricity, Forces, Energy, Particles, and the Universe) along with practical ideas, sample lesson plans, and lesson activities.

PLANNING FOR PROGRESSION IN SCIENCE

Paul Davies

INTRODUCTION

This chapter is presented in two interrelated sections: planning for progression and lesson planning. Science is unlike any other subject that children study at school. While in all subjects there will be things that children don't know or ideas that they do not understand, for science, children will have developed models, often unconsciously, about how the universe works. These ideas often make complete sense to the child and are 'internally consistent' with their views of how and why things work the way that they do. However, often these ideas will be in conflict with the scientific explanation, leaving the teacher and the child with the tricky task of unpicking and reshaping their thinking. Moving from a naive and simplistic personal view of science to the established and complex view is a journey and one we call progression.

   While this journey is, by its nature, a personal one (Asoko & Squires, 1998), the role of the teacher is to support pupils in their learning. Science is fascinating and supporting your pupils in deepening their understanding of this is the reason why many people become and remain science teachers. Progression and seeing the pupil as an individual are at the very heart of this. This chapter explores what research has to say about progression and how this has informed curriculum design. The chapter also looks at the challenges that pupils might face in their learning and encourages you to explore some of these ideas for yourself.

OBJECTIVES OF THE CHAPTER

At the end of this chapter you should be able to:

■ Describe the meaning of *pupil progression* in science education and give examples related to science
■ Understand some of the important theoretical perspectives associated with progression within teaching

DOI: 10.4324/9781003110187-9

- Understand how a science curriculum may be designed to support progression;
- Explore the key ideas about progression in science education and identify the complexities of supporting progression in learners
- Understand the underlining principles of curriculum design and lesson planning
- Be able to use curriculum materials to inform your own lesson planning
- Develop your skills at lesson planning to support progression
- Evaluate activities and approaches to teaching and learning which support pupil progression in science.

Describing progression

Children build up models of how they think the world works. Much of the job of a science teacher is to help them navigate shifts in their thinking from simple to complex ideas. In essence, this is what progression in school curriculum means. Take the following example about how a child might develop their understanding of the concept of 'what it means to be alive'. This is not an easy question to answer and something which biologists and philosophers of biology have, and still do, grapple with. However, imagine a young child exploring the world around them. Since the start of their life, living things – such as adults, possibly siblings and pets, plants in a garden or the home, food, and the natural world around them – have surrounded them. Research in this area has a long history, stretching back more than 100 years (Driver et al., 1985) and shows that children under the age of about 6 years tend to attribute the characteristics of life to anything that is active in some way or makes noise, for example a clock. As they get older, the idea of anything that moves becomes important to many children, meaning that most animals are seen to be living while plants are not. Up to the age of around ten years, the concept of self-propulsion becomes important, so things like rivers and the sun might be perceived to be living. As children become older, their views tend to change to regard only animals and, sometimes, plants as being alive. With this comes a view that there are specific characteristics, such as reproduction and respiration, which scientists search for when determining if something is alive or not. It is also here that children would see that scientists do not always agree, for example the debate about whether viruses are alive or not.

So what does this example tell us? One thing is that it says something about scientific ideas and how they increase in complexity. The initial ideas of the young child might be described as naive whereas the ideas of older children are more sophisticated. Also, the ideas move from everyday explanations, for example the hands on a clock appearing to move by themselves and the sun appearing to move across the sky, to scientific explanations, for example evidence from cellular biology. With this comes the importance of using evidence to support arguments too. So, the debate about the status of viruses requires knowledge about how living things work, the molecules that make up living things and the nature of replication.

In this example, we also see a shift in the nature and precision of language. So, 'anything that moves' is replaced with complex ideas about 'reproduction and respiration' with many layers of biological theory behind them. We also see how experiences that could be regarded as 'hands-on', such as pushing a toy train to see it moving, are replaced by experimental evidence, such as observing cells down a microscope. With this might also come mathematical modelling to explain patterns and probability, something that young children are incapable of doing.

The example also tells us about how learning happens. As a child grows up they experience the world in lots of different ways. An important one is the interactions they have with adults. The child whose parent is a professional biologist will receive a very different education about living things to one who is not. However, in UK schools, all children will have experienced the influence of science teachers; one of our jobs is to help pupils to develop their skills in – and knowledge about – science. Part of this education will involve exploring how biologists identify life and the complexities that surround these explanations. Another part of their role will be to encourage pupils to develop shifts in their thinking to match that of the scientific community.

Progression in the science curriculum then can be described as this journey from the everyday to the scientific and the naive to the complex. The aims of the National Curriculum have progression at their heart. These aims state that children should:

■ develop scientific knowledge and conceptual understanding through the specific disciplines of biology, chemistry and physics
■ develop understanding of the nature, processes and methods of science through different types of science enquiries that help them to answer scientific questions about the world around them
■ be equipped with the scientific knowledge required to understand the uses and implications of science, today and for the future

(UK Government, 2020).

These aims are about children being inculcated into the disciplines of science, with the curriculum as the tool to do this. However, whilst the aims are universal for all children, the journey that the individual child takes is very much a personal one. Learning in school is not a straightforward pathway but one that is often disjointed and disrupted. It is the role of the teacher to help pupils meet the aims of the curriculum and, in doing so, navigate their own journey in making increased sense of the world.

Task 8.1 **Describing progression in different science topics**

Pick a concept that is built up through the curriculum, for example 'energy', 'matter', or 'cells' and map the main stages of progression from Key Stage 1 to Key Stage 4. For example, for the idea of 'what is a living thing' the map could be:

- Some living things move, others don't
- All living things are made of cells
- All living things reproduce and respire
- All living things control their internal environment.

Discuss your maps with other student teachers.
    What patterns of increasing sophistication of ideas can you see?

Theorising progression

The basic premise of progression in learning science is easy to grasp: children develop more complex and sophisticated ideas as they move from everyday explanations to scientific explanation. The theory that helps explain this process is far from straightforward and has been the focus of much research. Braund (2008) proposes that progression should be thought of as developing *both* procedural and conceptual understanding. His ideas draw on the work of Gott and Duggan (1995) who argue that the cognitive processes that are required to solve problems and develop understanding have to be taught, rather than simply 'evolving'. Conceptual understanding focuses on knowledge and facts, something very important in science. It is also concerned with laws, theories, and models and explanations of how things work. Procedural understanding focuses on how scientific knowledge is constructed and includes the skills and processes of science, for example experimental design. These two ideas have powerful implications for teachers of science, particularly in England and Wales, where the curriculum is heavy with content as well as a need for children to develop an understanding about how the knowledge has been acquired.

The science curriculum is designed to support progression of both conceptual and procedural understanding. An important way that it does this is through the careful organisation of ideas so that they become increasingly more complex as a child moves through school. This design has not come about by chance or trial and error but has been heavily influenced by theoretical perspectives about how children develop knowledge and understanding. Much of this comes from the original work of Jean Piaget. The ideas of constructivism developed by Piaget, Vygotsky and others are covered in more detail in Chapter 10 and is very closely linked to progression in teaching and learning science.

Both Piaget's and Vygotsky's views of learning are about the child constructing it for themselves. This perspective of learning sees the child as an active participant in their own learning, as opposed to a passive recipient of information. This view of learning is described as social constructivist theory and has played a major role in shaping teaching and learning in the UK and many other countries (see Chapter 10). Nowhere has this influence been greater than in science education, something which can be attributed to two important research studies which have driven curriculum design and implementation in England and Wales. The first of these projects is the Children's Learning in Science Project (CLISP; Scott, 1987; STEM Library, 2020a). In brief, the project investigated children's ideas about scientific ideas and phenomena and revealed that often these are different from scientific explanations. The project yielded a wealth of information about how children think about scientific ideas; these potential 'starting points' of what a class of pupil may understand is an important one for teachers. For a detailed account of this project and its findings see Driver et al. (1985, 2014), but some examples of the ideas that children typically hold are:

- When solutes dissolve they disappear
- Air has no mass
- Electrical current is used up in a circuit
- Plants don't respire
- There are no products from combustion.

Since the CLISP project, extensive research has been carried out into children's ideas about science. A good review of this literature can be found in Gurel et al. (2015). The ideas that children hold, sometimes called prior conceptions, misconceptions, or alternative frameworks, can be very resistant to change. These terms are used to mean different things. Prior conceptions are ideas children have before they have learnt about the science. Misconceptions refer to errors that children have in their understanding, while alternative frameworks are ideas that children have that make sense to their way of understanding a particular idea but do not match the scientific explanation. Understanding what children think about scientific ideas and why is an important role of the science teacher.

The second research project which has had much influence on science education in UK schools is the Cognitive Acceleration in Science Education (CASE) project. This project drew heavily on the work of Piaget and Vygotsky in developing a programme which accelerated children's learning. CASE uses the important idea of Piaget that when confronted with new information, children either assimilate it into their current thinking or are forced to shift their thinking to accommodate a new model of explanation. From Vygotsky, CASE uses his ideas of the teacher supporting the learner to reach new understanding which they could not accomplish alone (see Hillier, 2018 for more information).

The CASE project led to the development of an entire course and extensive training programme for teachers where lessons were carefully designed to follow a pattern of learning unlike that seen in traditional classrooms. The lessons followed what was described as the 'five pillar teaching model'

(Adey et al., 1989). These pillars are show in Table 8.1. The CASE programme developed formal operational thinking in children across a range of scientific reasoning skills, including: classification, probability, control variable, and proportionality. CASE showed some impressive results in terms of how quickly pupils could move through the Piagetian stages. Not only was this seen in science but also in other subjects, including English and mathematics (Adey & Shayer, 2006).

■ **Table 8.1** 'Five pillar teaching model' from CASE (Adey et al., 1989)

Stage	Classroom activities
Concrete preparation	The teacher introduces the task and allows pupils to become familiar with the key ideas and vocabulary
Cognitive conflict	The teacher presents ideas which do not fit the children's ideas. Here the teacher helps the children negotiate the new ideas so that they can be accommodated into their models
Construction	The pupils work together to develop their new knowledge and problem solve
Metacognition	Having problem solved the children reflect on the strategies they used to do this
Bridging	The children are encouraged to see how their new knowledge can be applied in other settings and so become useful

Although social constructivist research has had a major influence on supporting progression in learning in science, other theoretical perspectives, which have become important for all school subjects, have become prominent in classrooms. A comprehensive account of many of these can be found in the companion guide to this book (Capel, Leask, Younie & Lawrence, 2022) but some significant ideas linked to supporting progressions are:

- Pupils use different ways to learn different concepts
- Pupils have differing levels of pupil motivation
- Pupils have different language skills
- Pupils come from different social backgrounds
- Pupils have different emotional responses to different topics.

Some of these ideas are explored in more detail in the Planning for Progression section later.

Design of the curriculum to support progression

Educational theory has sometimes been criticised for not being related to classroom practices. While this accusation may be true in some cases, much of the research into how children think about scientific ideas and learn in general has had an important influence on how school science curricula have been designed. Having explored some of the important theories related to progression in learning science, this section focuses on the influence that this research has had on curriculum design.

Piaget's theory of cognitive development is embedded in the Key Stage model that is seen in England and Wales where the curriculum has been more or less designed to try to match the ages of pupils (and their cognitive developmental phase) with the science content they are learning and assessment producers that are followed. Take for example the topic of Forces. This is introduced in Key Stage 1 as simple 'push and pulls' with children learning through play and embodied experiences, assessment following a pattern of teacher observation and pupils naming things. In Key Stage 2, the idea of friction is introduced with experiments used to test it. Still keeping to the observable world, learning develops with ideas about experimental design and controlling variables. In Key Stage 3, pupils are introduced to lots of new vocabulary and new ideas about pressure and moments. The learning here though is still based in real-world examples although the curriculum looks for links in learning both through the Key Stages and across subjects, for example in mathematics. Finally, at Key Stage 4 topics that require abstract thinking are introduced, such as the mathematics of forces and astronomy. These ideas are then developed in even more abstract ways in post-16 courses.

Designing the curriculum in this way has lots of benefits, the most important being that ideas are introduced when pupils are cognitively able to cope with them. It does ask two important questions though:

1 What about pupils with accelerated cognitive development? How does the curriculum cater for them? The answer to this question comes from how teachers support inclusion in their classroom (see later discussion).
2 Does the cognitive demand of the topics that are taught match the Key Stage where they are first introduced or developed?

This second question has been the subject of research for many years. The process of curriculum design is a complex and lengthy one, some of which is somewhat mysterious. Decisions have to be made about what topics are essential, the order in which topics should be taught (sequencing), and the depth of learning required. It would not be possible for curriculum designers to scrutinise every piece of research focused on cognitive demand of each topic, but it is hoped that this happens on some level. A very good piece of research that addressed these questions is well worth reading. It was written by Michael Shayer and Philip Adey in 1981 and provides theory to support the CASE project which followed soon afterwards. It also helped with the design of the National Curriculum. Task 8.2 encourages you to think about some of the questions raised in this research and how this influences your own teaching.

Task 8.2 **Exploring the cognitive demand of different science topics**

1 Pick a topic which interests you and examine the curriculum from Key Stages 1 to 4 to see how it is introduced and developed. You might find it useful to look at Best Evidence in Science Teaching resources (BEST, 2020, see Chapter 21)
2 Do some research into children's ideas about this topic and difficulties they have in understanding it (a good place to start is Driver et al., 1985; Asoko & Squires, 1998)
3 Identify any problems pupils might have in learning the topic material at either Key Stage 3 or 4.
4 Come up with two or three ideas that you think might help pupils make more sense of the problems you identified. Share these with your mentor as part of your lesson planning discussions.

The influence of both Piaget and Vygotsky in curriculum design also seen through how topics are revisited and become progressively more complex as a child moves through the Key Stages. Vygotsky's idea of learning with the 'more knowledgeable other' (often the teacher) where the child reflects on their current knowledge and understanding and is then presented with learning experiences to move them on in their learning is observed through what is called the 'spiral curriculum'. This term is derived from the influential work of Jerome Bruner who was interested in how revisiting ideas that have already been learnt is used as a foundation from 'layering' on new knowledge (Bruner, 1960). The role of the teacher here is to provide the 'scaffold' discussed earlier, to give the pupil enough support to make this a profitable experience. Too little support and the pupil fails too often, too much and the pupil can be too reliant on the teacher to help their learning and will be reluctant to take risks.

At the start of this chapter, I mentioned the idea that Braund has put forward that progress in science education is learning about the concepts or science and the processes of skills in science. He is not the only one to have considered this as a useful approach. The curriculum requires pupils to learn about a large body of information about biology, chemistry, and physics (the scientific concepts) and how these concepts have been developed and modified (the processes of science). We have seen that there is development in the complexity of scientific concepts in the curriculum and the same is observed in the process side of things too. At the moment, learning about processes of science is encapsulated in the curriculum through the idea of 'Working Scientifically'. Through this strand pupils begin by developing an understanding of how scientists collect data before going on to consider how they make sense of experimental evidence and, finally, how scientific ideas are debated and evaluated. Integrating these two strands of the curriculum has not been straightforward but Braund (2008) provides an important reminder that research has been carried out in these areas with some success. He points particularly to the work done by the Assessment of Performance Unit (APU) in science and a model they proposed (Qualter et al. 1990) which binds conceptual development in the spiral curriculum with a deepening understanding of the way that science is done. Braund argues this work – and Gott and Duggan's (1995) ideas about integrating learning about concepts and processes in science – are the best model of a coherent approach to support progress in learning science. As you gain experience in the classroom, you will realise that this is not always the easiest thing to accomplish. Task 8.3 encourages you to think about this further.

Task 8.3 **Investigating the spiral science curriculum**

In this task you will use the National Curriculum (or a curriculum appropriate for your context) and school teaching and learning materials, such as schemes of work, to map how a topic is developed from age 5 through to post-16 education.

1 Pick a topic which interests you and identify when it appears in each stage of your curriculum (Key Stage in the National Curriculum for England).
2 Record the ideas that pupils are expected to learn at each stage.
3 Compare what you have found to school schemes for work for ages 11–14 and 14–16 (English Key Stages 3 and 4). Look for links across them and how much extra learning is required.

Discuss your findings with an experienced teacher and ask them what issues they have found with the 'spiral curriculum'. These might include, for example, pupil assumptions that they have 'done this before' and challenges linked with Primary to Secondary School transfer.

LESSON PLANNING IN SCIENCE

As we saw earlier, the science curriculum in England has been designed, in some ways, in response to research into pupils' ideas about science and the best approaches for developing scientific concepts. This may make a teacher feel that they have little say in what pupils will be taught and when this will happen. Whilst this is true to some extent, teachers should not feel constrained by the curriculum but, instead, see it as a starting point for introducing pupils to the exciting world of science. Teachers have a lot of flexibility in how they design learning experiences and much freedom to use contexts that they – and their pupils – will find interesting. It is through the cycle of day-to-day planning that the curriculum comes alive. Becoming skilled at lesson planning takes time and is something a teacher has to engage with throughout their career. But the reward of seeing your pupils developing knowledge and understanding and those special 'wow' moments when pupils see the world through new eyes easily outweighs the hard work.

Planning for progression

A curriculum that is well designed to support progression still requires a skilled teacher to enact its aims. If there were a manual that told you how to do this, teaching would be a far easier job. But, teaching and learning involves complex social interactions; no two children are the same and neither are any two lessons. No matter how many times a teachers may teach the same topic to the same year group, the design of the lesson is always different, tailored to the pupils and modified through an iterative process of careful evaluation and reflection and design.

Thankfully, there are lots of resources, strategies, and a huge amount of professional knowledge that will support you in helping your pupils learn. The five that are most important when thinking about how best to design lessons are:

1 The National Curriculum
2 Examination syllabus material
3 Schemes of Work
4 Lesson planning strategies
5 Lesson evaluation.

and, in this section we consider each in turn.

The National Curriculum

The National Curriculum in England is a document which provides an outline of what children should learn across all the Key Stages (DfE, 2015). All local-authority-maintained schools in England must teach what is laid out in this document, with the vast majority of all types of school following it in some way, especially at Key Stage 4 (age 11–14). The National Curriculum was first introduced in 1989 and, since then, has undergone various iterations, the latest in science taking place in 2016 and 2017.

The National Curriculum is really a series of documents that break it down into frameworks for each Key Stage called 'programmes of study' for each subject (e.g. Science). The curriculum sets a series of aims that link well to the ideas of what a science education should involve, echoing very much the ideas of Gott and Guggan (1995) that Braund (2008) discusses. The aims are to ensure that all pupils:

■ Develop **scientific knowledge and conceptual understanding** through the specific disciplines of biology, chemistry, and physics
■ Develop understanding of the **nature, processes, and methods of science** through different types of science enquiries that help them to answer scientific questions about the world around them
■ Are equipped with the scientific knowledge required to understand the **uses and implications** of science, today and for the future.

The curriculum goes onto describe the content of what should be taught. This covers biology, chemistry, and physics (see Chapters 5-7), which address the first aim, as well as the Working Scientifically thread which addresses the second aim.

By setting out what material should be taught and when, the curriculum provides a rough timeline of how long it should take to cover all the material. It says nothing, however, about how the material should be taught or learnt – that is left up to schools and teachers to decide upon.

Examination syllabus material

If the material that is being taught is linked to a public examination at age 16 (e.g. GCSEs) or post-16 qualifications (e.g. A levels; BTEC qualifications, Highers), a syllabus produced by the specific examination board will be available. These documents are based on the National Curriculum (for that stage) and a shared set of broadly common criteria for post-16 courses. A syllabus – or specification – is a much more detailed document than the National Curriculum and provides a list of specific topics and sub topics that pupils must learn. The syllabus will give fairly clear guidance on how long should be spent on each topic and is often supplemented with lots of support materials to help with medium planning and individual lesson planning. There is plenty of flexibility for schools in how they decide to teach the material and, for courses where the assessment takes place at the end, in what order (sequencing) to teach the material. This leads us on to schemes of work.

Schemes of work

A scheme of work (SOW) is a document which provides, in considerable detail, the content of what should be taught, how long teaching should take (normally shown as lesson allocation or weeks), possible teaching and learning approaches, resources, opportunities for adaptive teaching, and assessment and links for cross curricular opportunities (including those that link to the Spiritual, Moral, Social, and Cultural aspects of the curriculum) as well as health and safety guidance and ideas about learning outside the classroom. SOWs can take multiple forms, including teacher guides, booklets that are produced for pupils and adapted and annotated by teachers, knowledge organisers, 'year on a page' structures, and so on. Examination boards often produce materials which can easily be edited and formed into a SOW. It takes a considerable amount of time to produce schemes of work and is often a task that is shared amongst the science team. SOWs should not be seen as concrete documents but, rather, organic, with teachers editing them as they progress through the course. It maybe that you are asked to contribute to the development of a SOW in your training schools. Task 8.4 will help you engage with using a scheme of work.

Task 8.4 **Exploring schemes of work**

Once you have located the schemes of work in your school, work through the following questions:

1 How is the scheme of work named?
2 What information is provided?
3 Is it a published scheme or developed in-house?
4 How is it organised?
5 What approach does it take (e.g. concept or context driven)?
6 What evidence is there that the scheme of work allows all pupils to access the learning?

Considerable effort goes into developing a scheme of work (SOW). The SOW should be designed to allow all pupils to progress. This means that the order that topics are taught is important and there should be opportunities for pupils to engage with the ideas in different ways and at different levels. The ordering of the topics to be studied is described as sequencing. How sequencing takes place for a particular topic will depend upon a number of things. One is what research has shown about the cognitive demand of the ideas (this links well with Task 8.2). Some of this research will be very academic, involving experiments and quasi-experiments with children. The outcome of the Best Evidence Science Teaching (BEST) research provides some very useful information on how to build up ideas (BEST, 2020) and is well worth exploring. Other research will come from the huge body of professional knowledge that exists in science education. This is something you will want to 'tap into' in your training. Before it was disbanded, the Qualifications and Curriculum Development Agency (QCDA) produced some excellent SOWs on all units in the National Curriculum that was being used up to 2006. These documents have been archived and can be found through the STEM Library (STEM Library, 2020b). They provide excellent examples showing how an idea can be developed across a Key Stage and beyond. For example, Key Stage 3 Unit 7A Cells is sequenced as follows:

■ Using microscopes to understand that living things are made of cells
■ Cell structure
■ Cell function
■ How new cells are formed.

As you explore SOW material during your training, pay attention to how specific ideas are built up. Remember that schools develop their own SOW; this gives teachers lots of freedom to design teaching and learning experiences that they think are the best for their particular pupils.

Lesson planning strategies

Having explored a scheme of work you will have seen that, whilst detailed, it does not explain precisely what will happen in each lesson. This is where the skill of a teacher really comes into play. Teachers plan lessons in all sorts of ways from incredibly detailed documents with minute-by-minute descriptions about what they want to do in the lesson, through to a series of ideas a teacher holds in their head. Of course, experience and confidence determine the approach to planning that a particular teacher might take, but as a beginning teacher, detailed planning is essential. Thorough lesson planning is very time consuming, but it means you are well placed to teach a good lesson and, most importantly, for the pupils to learn something. Poorly planned lessons are rarely effective and can be very stressful. Putting in the effort in the build-up to the lesson will pay off – so do this for yourself and your pupils. In addition to planning thoroughly, you also need to be aware of your subject knowledge. Science teachers often have to be generalists for some classes, for example teaching outside of their specialisation, but they also have to be aware of the difference between the science they learned during university courses and school science; these things are different. It might be tempting to teach pupils everything you know about a particular topic which was the focus of your undergraduate degree project – this is mistake. Not only will the material be, almost certainly, too challenging, it will not be what pupils should actually be learning. That is not to say you shouldn't bring your expert knowledge and scientific interests into the classroom, of course these are good things, but you need to make sure you have secure subject knowledge of the topic you are teaching. This includes the level of detail, how the ideas link to others across the curriculum, and the language that the school science curriculum uses to describe and explain the specific ideas.

Another important consideration is the amount of information you want the pupils to encounter in an individual lesson. Recently, research in Cognitive Load Theory (Sweller, 2011) (see Chapter 9) has been important in many schools and this can be used as a helpful guide when thinking about how to present new information. The theory brings together ideas about working memory (explored in more detail in the *Supporting pupil progress* section) and schemas. The key idea of the theory is that new information is first of all used in the working memory (Gathercole & Alloway, 2008). For learning to take place this information needs to be transferred to long-term memory where it is assimilated into pre-existing schemas (best thought of as groups of information focused on a common theme). In lesson planning it is important to consider how many pieces of information you expect pupils to use at one time (cognitive load). Research suggests that pupils can hold between three and seven pieces of information in their working memory at one time. If too much information is presented at once, pupils will not be able to use it effectively and learning will be disrupted. Richard Mayer provides a very useful guide that will help you plan for reducing cognitive load (see Mayer, 2002).

Planning lessons is hard work and every lesson is different. To help you in your early stages of planning it might be useful to consider the structure of a typical lesson:

■ A starter: this sets the scene of the lesson, maybe makes links to previous learning and helps pupils to understand 'what the lesson is about' (i.e. reveals the learning outcomes to them). This section of the lesson might also be used to help you understand what the pupils already know about the topic

■ Main activities: episodes where the pupils are actively engaged in their learning. This could involve listening and observing, working in groups, working individually, carrying out practical work, role-play or class discussion

■ Mini-plenaries: these short episodes take place after each main activity which sum up what the activity was covering and allow the teacher and pupils to gather information about what they have learned from the activity. These findings should feed into subsequent activities

■ A main plenary: normally comes at the end of the lesson to allow for further teacher and self-assessment and a review of learning. There might also be opportunities for self-reflection on personal learning

In order to plan lessons you first need to consider what it is you actually want the pupils to learn.

These ideas will be in the scheme of work, often as a list of ideas or topics that have been distilled into *learning objectives* for individual lessons or units. Learning objectives are lists of things you want pupils to have learned. For example, in a lesson on photosynthesis the learning objective might be *pupils will learn about how the structure of the leaf is related to its function in photosynthesis*. That is a fairly broad idea but also one which is hard to assess, that is, it is not straightforward for a teacher to know if and when pupils meet this objective. Therefore, in the classroom, learning objectives are normally transformed in *learning outcomes* (LOs). These are things that you expect the pupils to be able to demonstrate in order to show they have met the learning objective. Outcomes are normally activities which are measurable, e.g. being able to describe, explain, calculate, evaluate, and so on[1] (notice that the LOs all include verbs). For a photosynthesis lesson the learning outcomes might be:

1 Know that photosynthesis mainly takes place in palisade cells
2 Describe how the leaf is adapted for photosynthesis
3 Explain how carbon dioxide reaches the chloroplasts
4 Suggest how plants without leaves carry out photosynthesis.

The LOs in the previous example become progressively more challenging; this is something you will often see in lessons. An important part of your planning to ensure progress by all pupils is the notion that the pupil should be able to access the curriculum in different ways and at different levels; this is called adaptive teaching. Pupils will access the curriculum is all sorts of ways; this might depend upon their prior attainment, if they have English as an additional language (EAL), if they have lower than average reading and writing speeds, or if they have other specific educational needs or disabilities (SEND; for further information see Chapter 18).

Different schools have different approaches to how learning outcomes should be developed. A common approach is using learning taxonomies. These are tools that will help you design LOs and think about adaptive teaching. There are lots of different learning taxonomies but two found in schools are: Bloom's Taxonomy and Structure of the Observed Learning Outcome (SOLO). Whilst in educational academic literature there is lots of debate of about how – or even if – these taxonomies should be used in the classroom, you may find them useful. Both of these taxonomies take a hierarchical approach to learning, which sees activities such as 'knowing' and 'identifying' as more simple than 'analysing' and 'evaluating'. Task 8.5 may help you reflect on the potential advantages and disadvantages of using learning taxonomies in your own lesson planning.

Task 8.5 **Lesson outcomes**

For a lesson that you have observed, consider the following questions:

1 What did the teacher intend the learning outcomes to be?
2 How was the lesson adapted to enable all pupils to access the learning outcomes?
3 How did the teacher share the learning outcomes with the class?
4 How did the teacher know that the learning outcomes had been met by the pupils?

Different pupils will respond to learning outcomes in different ways. This will be for all sorts of reasons, such as motivation and interest, prior knowledge about the subject, and even things like their mood and how near the lesson is to lunchtime. Taking the photosynthesis example from earlier, for LO2, one pupil may give a fairly simple answer along the lines of: 'Leaves are thin and green, this means gases can get in and out easily and there is lots of pigment'. Another pupil might respond with something like: 'Leaves are thin and so there is a short diffusion pathway for carbon dioxide and water to reach the palisade layer. Palisade cells are packed with chloroplasts and located near the top of the leaf. The chloroplasts contain chlorophyll, a green pigment which absorbs sunlight'. The second response meets LO2 at a higher level than the first, showing that LOs can be meet at different levels, or, put another way, progression can take place within each LO. Chapter 18 explores how teaching can be adapted to scaffold learning so that all pupils can access the lesson outcomes.

Once you have decided what the learning objectives of the lesson will be, it is time to start thinking about the details of the lesson and the resources you will use. There are hundreds of resources available for teachers and your training school will have many. However, deciding on which to use and what approaches to take in the lesson isn't easy. For example, you might want to take an investigative approach in your lesson, or you might want to make the lesson set in a very real context, or perhaps you want the pupils to listen to you explaining before they do something with this information. To decide on which approach to use and its associated resources, it is useful to watch experienced teachers and see what they do. Talking your ideas through with colleagues in your department will be useful, as will sharing ideas with other student teachers.

Whatever approach and resources you are going to use, the following questions will help you think about how the lesson will be enacted:

■ Why do I want to use this approach?
■ Will my resources support this approach?
■ What will the pupils learn through this activity?
■ Will all the pupils be able to access the activity?
■ How will I support individuals and different groups of pupils?
■ How can the activity be extended?
■ How long will the activity take?
■ How will I organise the activity?
■ How will I assess if the pupils have learned what I intended them to learn?

In order to help plan the lesson it is useful to consider these questions:

1 What is it that the pupils know and understand at the start of the topic?
2 What do I want them to know and understand at the end of the lesson?
3 What sequence of activities is going to support pupils in their learning?
4 How will I know when the pupils have learned what I want them to?

Eliciting what pupils know and understand at the start of the topic

As research and experience show, pupils come to the lesson with all sorts of prior knowledge about the topic you are going to be teaching about. Understanding what is *knowledge* is an important aspect of a lesson. It might be that a pupil already knows and understands much about what you are going to cover in the lesson; in this case they will be bored with no opportunity to progress in their learning. Likewise, a pupil might have alternative frameworks about the topic, in which case you are going to need to do things which help the pupil understand what these are and change their ideas to match the scientific explanation. There are lots of things you can do as part of your lesson planning to better understand what pupils might know and, equally, there are different things you can do during the lesson to help pupils reveal this to you. Activities that take place in the lesson to elicit pupil knowledge and understanding are called diagnostic tasks.

Some useful things to do are:

1 Find out what the class should have been taught about the topic already. This might be from different Key Stages or in other subjects. Look at the National Curriculum and talk with colleagues at your school to find out about these things

2 Talk with colleagues in the science department and learn about their experiences of how pupils think about the topic and what they typically know

3 Consult the academic literature to find out what is known about the ideas that children have about the topic. Good places to start are Driver et al. (1985, 2014)

4 In the lesson, use diagnostic tools to elicit prior knowledge. These might include activities such as:

  a Formative questioning

  b Multiple choice questions where the distractors are common misconceptions

  c Refutation tasks where pupils have to explain why a misconception is wrong

  d Using statements which pupils have to decide are true or false

  e Concept cartoons which explore common misconceptions.

  f Argumentation activities which encourage pupils to defend their ideas (Osborne, et al. 2016)

Task 8.6 **Eliciting pupils' ideas about science**

Pick a science topic which you find challenging and explore this with a class in a lesson or, if possible, with a small group of pupils outside of the lesson.

1 Using one of the strategies (a-f) outlined above, devise an activity which explores what pupils know and understand about the topic you have chosen. You might need to consult some academic literature on children's ideas about science (see Driver et al., 1985, 2014)

2 Record what you find out. How does it compare with the literature?

3 Devise an activity which you think would help pupils to move from their current thinking to a more complex scientific explanation.

4 If possible, try this activity out with a class or group of pupils. Evaluate how effective it was at supporting their learning.

Supporting pupil progress

Your diagnostic assessments in the classroom will reveal that there are a range of ideas and knowledge about the topic amongst the pupils. It is not easy catering to all these ideas and it can seem daunting. It is impossible to design individual lessons for each pupil so, instead, you need to think about designing a range of learning activities which, through the lesson and series of lessons, support all of the pupils in making progress.

Schools are full of resources and you should familiarise yourself with what materials can be found in yours. You should consider how a particular approach will help your pupils to understand the concept or skill you are teaching. For example, abstract ideas can come to life with the help of a demonstration while the development of mathematic skills is often made more accessible when pupils use data that they have generated for themselves.

You also need to think about how you will ensure individual pupils can make progress for themselves. As noted earlier, your learning outcomes should be designed to allow all pupils to access the curriculum and make progress in their learning. Schools collect a large amount of data on their pupils and this can feel a little overwhelming at first. However, you should consult this information and talk with teachers in your department to better understand what it means. The kinds of data that will be most important to you include:

- Pupils' prior attainment in tests, examinations and in predictive assessment (e.g. MiDYIS, see CEM, 2020)
- Whether or not the pupil is recognised as having English as an additional language (EAL)
- Specific learning needs of each pupil

In order to help you plan, you should think about how you can adapt your teaching in order to help all pupils make progress. Chapter 18 offers some useful insights here. There are lots of different ideas which you will need to experiment with and discuss with other teachers in your school, but some useful, straightforward ideas to consider and try are:

1 Use questioning that is targeted to ensure pupils are challenged but not overwhelmed
2 Try using a resource, for example a worksheet, which has different levels of challenge. This might be a single worksheet that pupils start at different places (see concept of ramping, chapter 16)
3 Think about how you group the pupils: will you place pupils with a similar understanding of the topic together or is it better to mix them up?
4 Think about roles that pupils might take in collaborative tasks.

An important aspect of supporting progression is how you sequence and divide up the ideas and concepts you want the pupils to learn. As mentioned earlier, research in working memory has become important in school. When presenting new information you should consider how much you introduce at one time. The best teachers tend to introduce a few ideas and then give pupils the opportunity to practice the material. You need to be mindful about how much teacher-talk you use as well. The quality of what you say and show pupils is more important than the quantity. So, you might introduce a concept and model it on a white-board or computer presentation programme and then set a task where the pupil use these new ideas. You could then use questioning and follow-up explanations to reinforce the idea and address problems pupils may have. Research has also shown that pupils learn well when the teacher models how to complete tasks (Schwarz et al., 2009; Rosenshine, 2012). Making explicit what you expect from the pupils is very important here. Sometimes you might share 'success criteria' with them or show them a piece of completed work so that they understand what to aim towards. Talking aloud as you explain how you tackle a question or problem is also useful as this reveals to the pupils the thought processes that you go through in your own learning.

Assessing pupil progress

Assessing pupils, both during a lesson and through work they produce that you mark away from the lesson, are key for successful pupil progression in learning. Within a lesson teachers use lots of different ways to find out if their pupils understand and can use the ideas they are being taught. A comprehensive dis-cussion of how to use assessment to promote learning can be found in Chapter 16, but a key tool at your disposal will be questioning. Teachers use questions all the time to find out if pupils know what to do in a particular task and what they understand. You might ask a very closed question, such as 'What is the symbol for hydrogen?' This question has only one correct answer and tells you some-thing about pupils' ability to recall information. Open questions tend to elicit much richer responses and encourage pupils to elaborate on their understand-ing or make links between topics. Another useful approach is to ask pupils to explain how they worked out a particular answer, something which reveals their thinking and problem-solving approaches. You might also ask pupils to defend their position or build on the answers of another pupil. All of these approaches help you determine what the pupils know and can do and, importantly, what they don't know and can't do.

It can be tempting to plough on through your lesson. This is understandable because you will have spent time planning it and perhaps preparing resources, but don't be tempted to rush through everything you have planned if the pupils are struggling to understand the content in the early part of your lesson. If your assessment strategies reveal that pupils have not understood the material you have been teaching, you need to address this. This might mean taking a different approach and explaining the ideas in a different way. Don't feel bad about this but seize opportunities like this to show pupils that learning is hard, it takes effort and resolve, and different people think about the same idea in different ways.

Beyond questioning, typically in a lesson pupils will produce work which will tell you much about their knowledge and understanding and the progress they have made in their learning. However, some of the evidence you will collect is less concrete. This will include, for example, things they say, how they work with others, their ability to manipulate apparatus, and how they work through a problem. It is hard to capture these types of data and they are probably not things that you would record in a mark book, but they provide you with important information about each of your pupils.

As you develop your repertoire of assessment approaches, you might find it useful to try the following things:

1 Focus on two or three pupils in each class (perhaps of differing levels of prior attainment in science) and make a note of the type of evidence outlined earlier. This will help sharpen your attention to pick up on the nuances of how pupils reveal their knowledge, understanding, and abilities

2 Encourage the pupils to develop a learning log – this is a diary where pupils record their ideas and reflections about what and how they have learned in each lesson. There are lots of examples on the internet, but you will find that these reveal much about what your pupils think and how they think. Chapter 5 explores some of these in biology. Above all else, they will often surprise you and help you develop as a teacher much more fruitfully than many other approaches you will encounter.

Some of the learning that pupils complete happens away from the classroom. This is especially important for post-16 courses where pupils often have private study periods. Planning for how pupils should use these opportunities is important. You should find out if your school has a homework policy and what types of homework tasks it recommends. It might be that the schemes of work include information about homework but you need to think carefully about how to use homework to support effective learning. Don't be tempted to use homework tasks as a way to 'complete the work started in the lesson'. Instead, design tasks which encourage pupils to use their learning, demonstrate skills or knowledge and understanding, or possibly prepare for the next lesson (so-called flipped learning). Useful guidance on homework will come from talking with teachers in your school but the following will provide a framework for your planning of suitable tasks.

Effective homework tasks:

■ Test pupils' knowledge and understanding or skills
■ Allow pupils to build on their learning
■ Allow pupils to practise and develop mastery
■ Help pupils make links in their learning
■ Encourage pupils to share their learning with their family and others
■ Inspire pupils to 'want to know more'
■ Help prepare pupils for subsequent lessons
■ Shouldn't take longer than the time allocated by the school.

Lesson evaluation

As discussed in Chapter 2, one sure way of improving your lessons is to reflect on how they went and evaluate their effectiveness. Reflection is an important aspect of being a teacher. If you ask pupils about your 'performance' in the class they can often give useful feedback about the types of task they enjoy and if they understood what they are meant to be learning but they find it hard to analyse why a lesson or particular activity was helpful in them making progress. To become a reflective teacher takes practice. In the early part of your training you should talk with teachers whose lessons you observe to see how they 'unpick' what happened. When you evaluate your lessons don't be too hard on yourself. Even very experienced teachers have lessons which go wrong; the important thing is to learn from the lesson to work out what is effective and what you might do differently in the future.

SUMMARY AND KEY POINTS

This chapter has explored progression in science education and focused on:

- Ideas about pupil progression in education
- Some of the important theories that have informed what we know about how children learn and understand science
- How these theories have been used to shape science curricula
- Lesson planning for progression
- Lesson planning to help elicit what pupils think about scientific ideas and strategies to address these
- Ideas about assessment and how this is linked to effective lesson planning
- Ideas that support all pupils in making progress in your classroom.

Check which requirements for your initial teacher education you have addressed through this chapter.

NOTE

1 In some schools, it is common practice for pupils to write down the learning outcomes at the start of the lesson.

FURTHER RESOURCES

Alonzo, A. C., & Gotwals, A. W. (Eds.). (2012). *Learning progressions in science: Current challenges and future directions*. Springer Science & Business Media.

This book develops an argument about the need to design curricula that promote progression in learning science. Developed from a conference focused on this issue, the book brings together some of the most important researchers in this field and provides an accessible way to engage with the academic literature.

BEST. (2020). *Best evidence science teaching*. University of York. https://www.stem.org.uk/best-evidence-science-teaching.

Best Evidence Science Teaching (BEST) is a collection of free research evidence-informed resources for effective teaching of difficult ideas for Key Stage 3. The resource contains materials helpful in sequencing ideas and assessing pupil learning.

Hillier, J. (2018). Planning for pupil understanding and learning behaviour. In I. Banner & J. Hillier (Eds.), *ASE guide to secondary science education* (pp. 89–98). ASE.

Written for teachers, this chapter helps teachers understand how to design lessons that support all learners. Drawing on important research, reading this will help you develop an overview of current thinking about how to best support effective learning in the science classroom.

Rosenshine, B. (2012). Principles of instruction: Research-based strategies that all teachers should know. *American Educator, 36*, 12.

Freely available online, this document will be invaluable as a guide to help you design lessons which really help pupils learn. Evidenced-based, it provides a series of clear instructions about what works in the classroom and why. Using these ideas will really support your early planning, helping ensure your lessons are well-crafted and provide pupils with opportunities to learn.

Some useful websites with resources to support diagnostic assessment through multiple choice questions with distractors, refutation tasks, and argumentation activities include:

- http://www.iop.org/education/educate/page_67488.html;
- http://assessment.aaas.org/topics;
- https://diagnosticquestions.com/RSC;
- https://www.york.ac.uk/education/research/uyseg/projects/developingdiagnosticassessments/
- https://thescienceteacher.co.uk/balancing-chemical-equations/

THE SCIENCE OF LEARNING

Nick Pointer

INTRODUCTION

As a teacher, whether you are just starting out or have ten years' experience, the learning process can seem anything but straightforward:

- Pupils appear to understand complex scientific ideas one day, yet a few weeks later in an end-of-unit test perform as if they'd never attended that lesson
- When following a seemingly simple series of instructions, pupils mix up or forget steps for no apparent reason
- Having all listened to the same explanation of a new concept, some pupils 'get it' right away, others develop a range of misconceptions, and other fail to understand or learn at all.

The aim of this chapter is to explore some of these common classroom events by examining how pupils learn.

Learning has been defined in many ways over the years; for example, a historical definition from psychology might be *a long-lasting change in behaviour as a result of experience*. On the other hand, advances in neuroscience have long promised to help us understand the biology of learning in more detail. Looking at the structure and function of the nervous system with a particular focus on the brain, neuroimaging technologies such as fMRI now allow the mapping of neural activity in real time, and there have been some attempts at drawing out the implications of neuroscience findings for classroom teachers (e.g. Howard-Jones et al., 2018). Neuroimaging studies can also add significant weight to the testing and validation of psychological theories of learning – for example, fMRI studies have provided evidence that differences in the phonological processing system are associated with developmental dyslexia (Pugh et al., 2000).

DOI: 10.4324/9781003110187-10

Yet, even with the latest developments we cannot look at a neuroimage and conclude that 'Sam is thinking about whales right now but has developed the misconception they are fish'. As a result, Didau and Rose suggest that 'neuroscience is probably the wrong level of description to provide meaningful insight into classroom practice' (2016, p. 247) and suggest that a definition of learning that is more helpful for classroom teachers comes from cognitive psychology – learning involves a *persistent change in long-term memory* (Kirschner et al., 2006).

For teachers, the field of cognitive psychology offers robust empirical evidence about how people learn and well-tested theories explaining the processes involved. This 'science of learning' can help you as a student teacher to answer the question: *How can I best support my pupils to learn?*

OBJECTIVES

At the end of this chapter you should be able to:

■ Understand a simple model of how the mind learns new information
■ Apply insights from cognitive psychology to the implications for instruction in the classroom
■ Support your pupils to consolidate their learning effectively over time.

A SIMPLE MODEL OF MEMORY

In science we use models to explain and predict natural phenomena, and we acknowledge that they are not necessarily accurate representations of reality. All models are 'wrong' to some degree – they are simplified representations of complex or opaque systems – but nevertheless remain useful in their predictive power. See chapter 11 on 'Models and Modelling in Science Teaching' for more on this. A model which can help illustrate the cognitive processes involved in learning is shown in Figure 9.1, which consolidates and simplifies the models developed in cognitive psychology over the past 50 years. It is important to note that it does not seek to depict the physical structures of the brain but rather to explain how the process of learning appears to take place.

Before looking at the inner workings of the different parts of this model, we will first simply consider how they interact in the learning process, which begins with the 'environment'. This comprises any stimuli that our senses are exposed to; for a pupil these might include the sound of your voice, an equation written on the whiteboard, or the sight of their friend making a face across the room. Whilst we like to think we decide what to pay attention to, 'our minds have their own wishes and desires when it comes to the focus of attention' (Willingham, 2009, p. 208), as demonstrated when a door slams at the back of a room, rapidly and automatically redirecting pupils' attention.

Our senses are exposed to more stimuli than we can reasonably attend to; the information that you do attend to is processed in 'working memory'. Working

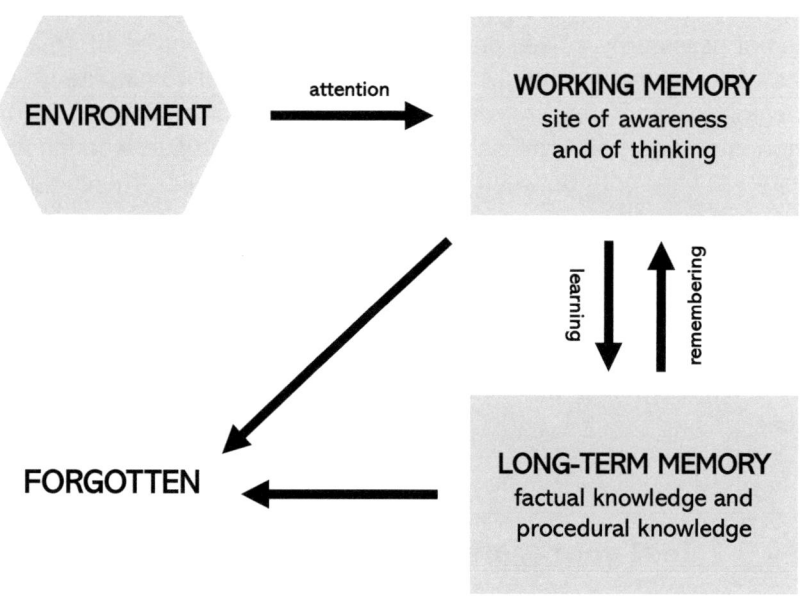

memory can be thought of as the part of the mind which deals with 'the stuff you're thinking about' (Willingham, 2009, p. 14). Processing new information in working memory can allow it to be transferred into 'long-term memory' for storage. This is necessary, since the capacity of working memory to process new information is very small, whilst the storage capacity of long-term memory is essentially limitless. If you think hard about something in terms of its *meaning* (and how it relates to existing knowledge you have), it is more likely to be transferred into long-term memory, i.e. learnt (Craik & Lockhart, 1972).

Information is organised and stored in your long-term memory, from memories as recent as 30 seconds old, to knowledge acquired years previously. Recalling information into working memory strengthens its place in long-term memory; knowledge that lies unused will become harder to retrieve over time until you cannot recall it at all.

Before considering the implications of this for your classroom practice, we must explore the two main components of this simple model of the mind in more detail.

The architecture of working memory

In everyday language you sometimes hear people refer to 'short-term memory' as a temporary store for new information. However, cognitive psychologists instead refer to 'working memory' as the part of memory involved in *actively processing* new information, with a highly finite capacity to do so (Cowan, 2010).

This capacity varies between individuals and increases as we mature, plateauing in adolescence and decreasing again in old age. Whilst current estimates for its finite capacity have settled around three to five pieces of information, this is not necessarily a fixed property in practice; for simpler information like strings of characters you might be able to retain closer to nine items, but for more complex, interacting concepts this can be much lower (Cowan, 2010). Thus, any notion of a precise 'size' isn't particularly meaningful, as working memory capacity will vary both between learners and depending on the complexity of the material in hand.

It is very easy to exceed the capacity of working memory: if you attempt to think about too much information simultaneously then your working memory will be overloaded, and some of the information will be lost. This might not be initially obvious – it may still be possible to ostensibly complete a task, whilst failing to learn anything from the process. Task 9.1 is a chance to test this out yourself.

Task 9.1 **Test your working memory – Experiment 1**

Below is a sequence of 15 characters. Give yourself *five seconds* to study the list (get someone to time you if necessary) and attempt to remember them all, in order. Then close this book and write down as many as you can.

O62 HC1 62O HO2 2CO

How did you do? This task required you to hold the characters in your working memory. Given the limitations of working memory, almost no one will remember all the characters.

However, it is not all bad news. To reduce the load imposed on working memory, information can be recalled from long-term memory to make sense of and simplify new information from the environment, in a process called 'chunking'.

Take care: chunking is sometimes used in an instructional context to refer to 'breaking tasks down into small steps', but in a cognitive setting we instead take it to mean the process where prior knowledge from long-term memory is used to reduce the load on working memory. This is best illustrated by an example, in Task 9.2.

Task 9.2 **Test your working memory – Experiment 2**

Below is another sequence of 15 characters. Give yourself *five seconds* to study the list and attempt to remember them all, in order. Then close this book and write down as many as you can.

CO2 H2O C6H12O6 O2

How did you do? You almost certainly remembered more than previously, which may seem surprising since these are the same characters you saw in Task 9.1.

This time you no longer had to memorise 15 individual units: you only had to remember four 'chunks' of chemical formulae. You might even have chunked the sequence further: if you noticed that this was simply the unbalanced symbol equation for photosynthesis then you may have been able to store this as just one unit to remember.

In Task 9.2, your existing *knowledge* of chemical formulae supported your working memory. To explain this, we need to explore how knowledge is organised in long-term memory.

The architecture of long-term memory

In everyday terms, people sometimes refer to our long-term memory like a computer hard drive. Previously learnt information is neatly stored and categorised in folders, and to access our 'files' we simply type in the relevant keyword or category and off we go. If we cannot recall a certain memory, it might be corrupted, we might have run out of storage space, or perhaps there was a filing error.

However, cognitive psychology has found this to be a poor model. The capacity of long-term memory appears to be practically limitless, in stark contrast to the highly finite limits of working memory. The organisational nature of the 'filing system' model is also flawed. Cognitive psychologists believe that stored memories are not neatly arranged like entries in a database; instead they are organised in messy, interconnected webs of knowledge. We call these 'schemas' or 'mental models', and they are what we use to make sense of and interpret new information.

Many of our schemas organise knowledge semantically, i.e. according to some associated *meaning*. For example, you will have a schema for 'trees', which might include memories of a tree in your childhood garden, the word for 'tree' in French, or the biological knowledge you have about plant structures. These schemas can vary tremendously in complexity, and well-connected smaller schemas can combine to form increasingly large mental models, such as those you might have for 'atoms', 'electricity', or even 'science' (for a more detailed overview see Jones et al., 2015).

Gatekeepers to learning

The interplay between these components of the mind can lead to learning, yet there are potential bottlenecks to the process, such as the role of attention. With such an immeasurable array of stimuli flooding into our senses, our brains must constantly filter through this information to attend to a limited number of things at any one time.

This has stark implications – if we don't think about something, then it is unlikely for it to be transferred to long-term memory. In the classroom, it can be tempting to try to direct pupils' attention by stimulating their interest. For example, beginning a unit on the rock cycle by using chocolate bars to model the melting and solidification of rocks might seem to represent an exciting and memorable lesson, but what are pupils thinking about, and what will they be able to remember? They will probably remember using the Bunsen burners, getting melted chocolate all over their books, and perhaps a few stolen bites of confectionary, but it's less likely they will remember the difference between the cycles undergone by igneous or metamorphic rocks. As Willingham states:

> *'Students remember what they think about'*
>
> *(Willingham, 2009, p. 79)*

The limited capacity of working memory is another *gatekeeper to learning*, and research into how teachers can account for its finite nature has led to the development of cognitive load theory (CLT). By optimising the demands placed on working memory (the 'cognitive load'), CLT considers how to best organise instructional approaches to support learners to think hard about the materials they study, whilst avoiding overloading their working memory capacities (Sweller et al., 2019).

IMPLICATIONS FOR THE CLASSROOM

Implications for the classroom: expertise and breaking down knowledge

As you gain more knowledge, organised in long-term memory as schemas, you gain more *expertise* in a particular subject or topic. However, the mental models of beginners are not just more basic versions of the more advanced understanding of an expert; rather, they are fundamentally different. This means that novices and experts perceive and solve the same problems differently. Novices tend to focus on more superficial, surface-level features of problems, whilst experts are more likely to notice their deep, underlying structural aspects (Chi et al., 1981).

What appears to be greater expertise is therefore often the automated use of these schemas to access and interpret new information. If you were to read an extract about the industrial processing of iron ore, your knowledge of 'mixtures and compounds', 'displacement reactions', and 'reduction' would be easily drawn on to help you make sense of it. On the other hand, novice pupils are less likely to be able to filter and recognise the most relevant information to attend to. So, when reading the same extract they might allocate valuable working memory capacity to thinking about the transporting of the ore, the sizzling furnaces, or trying not to forget the chemical symbol for iron, rather than displacement reactions.

Additionally, without well-developed schemas, beginners are unable to 'chunk' as effectively and so must try to process large quantities of new information simultaneously, increasing the likelihood of *cognitive overload* of working memory. In the classroom, the consequences of this can range from place-keeping errors (forgetting the place of a step in a sequence of instructions) to incomplete recall, failing to follow instructions accurately, or even task abandonment as learners give up in frustration (Gathercole & Alloway, 2004).

So, whilst two people might have similar limits on their working memory capacities, the extent to which they can process and make meaning of new information (and by extension, learn) can differ wildly as a function of the prior knowledge they have in long-term memory. Whenever pupils encounter a topic where they have little prior knowledge, their working memories are easily overloaded with too much new information. To prevent this from happening, you should break-down new learning into accessible, bite-sized chunks.

Unfortunately, when you know a lot about a particular topic, it can be extremely hard to break down what you know into the component parts which might help a pupil encountering it for the first time. Heath and Heath (2007) refer to this as the 'curse of knowledge' – the natural and unavoidable phenomenon where your memories of the learning process become more automated as proficiency develops. As a result, you must be wary of your *assumptions* about pupils' existing knowledge, skills, or experiences you might be able to build on in your teaching.

This next task is an opportunity to practise breaking down your knowledge. This will allow you to gradually build pupils' expertise over time, ensuring that they have the opportunity to be taught and to practise each individual component *before* you ask them to put them all together.

Task 9.3 **Breaking down knowledge**

Consider an upcoming sequence of learning that aims to develop pupils' abilities in a specific area. For example, I might have a sequence of learning that aims to support pupils to *Understand the flow of electric current*.
  Using your sequence of learning:

1 Write or find an 'end-goal task' that might sit at the end of this sequence. An exemplar end-goal task might be: *Describe why electric current flows through metals but not through plastics*. Write out an excellent model answer to your end-goal task, in full.
2 Annotate your model answer with all the knowledge that underpins it. Remember the 'curse of knowledge' and take care to avoid making assumptions – the more detailed and specific you can be, the better. In my example, my annotations might include:
   ■ *I know that atoms have an 'atomic' structure, with positive protons and neutral neutrons forming a central nucleus, orbited by negatively charged electrons.*

■ *I know that neutral atoms have equal numbers of protons and electrons, whereas ions have differing numbers of electrons or no electrons associated with the nucleus.*

■ *I know that metallic materials have a regular structure of fixed positive ions, surrounded by a sea of delocalised negative electrons.*

■ *I know that electric current is the rate of flow of electric charge from positive to negative.*

3   Having produced your final list of the requisite knowledge, consider: will pupils have been explicitly taught all of this a) before starting this sequence, in preceding lessons, or b) by the end of the sequence itself? If the answer is neither, consider how to build this into your planning, as in step 4.

4   Take your full list of requisite knowledge and sequence it in order. Aim to build from simple concepts to more complex ones, and from concrete contexts to more abstract ones. Finally, break this into accessible, lesson-sized units and map them across the time available for your teaching sequence.

This is a suitable task to collaboratively complete with your mentor, who in particular might be able to help you break down the foundational knowledge that underpins your model answer in a detailed and explicit manner.

Implications for the classroom: the multi-component model of working memory

A more elaborate model of working memory divides it into two separate components: one that deals with non-verbal imagery (the 'visuospatial sketchpad') and another that handles words and audible information (the 'phonological loop') (Baddeley, 2012).

Reader beware: this is often confused with Paivio's *dual coding theory* (1971), a different model that seeks to explain how information is encoded in memory, which happens to also feature a visual and verbal pathway for information processing. This should also not be confused with *learning styles*. The idea that there are 'visual learners' or 'verbal learners' has been repeatedly debunked, with evidence indicating that all people think using both pathways (Dekker et al., 2012).

In Baddeley's model, each separate component has its own processing capacity. However, providing pupils with multiple inputs requiring the same component (e.g. asking them to read sections of text whilst providing a verbal explanation) leads to cognitive overload. This is the *redundancy effect*: in this example, reading occupies the phonological loop as the words are converted into phonemes in pupils' heads, as does narrating out loud, and so their working memories will become overloaded as they cannot process both concurrently (Sweller et al., 2019). Instead, consider how to best complement an image or diagram with a verbal explanation, or allow pupils reading time for a worksheet or slide before beginning to read or explain it.

You can also consider how to reduce *extraneous* sources of cognitive load – those that originate from how information is presented, rather than the complexity of the material. Attention switching is one example of this. Presenting pupils with a diagram with a key of labels at the side, for example, requires them to rapidly switch their attention between the key and the marked points on the diagram. This is the *split-attention effect* and has been shown to reduce the available capacity of working memory (Sweller et al., 2019). Simply embedding the labels in the diagram itself can avoid this negative consequence.

Implications for the classroom: instructional approaches

In the classroom, it might seem intuitively beneficial to use practical activities to guide instruction; for example, introducing osmosis by asking pupils to make observations about changes in mass of potato pieces. In practice, however, without the prerequisite, well-developed schemas in long-term memory, novices are forced to try to process all of the information involved simultaneously.

This leaves little to no working memory capacity available to think about the content matter involved. As a result, teaching new content through minimally guided approaches – allowing pupils to explore new information and make sense of it themselves or using problem-solving to develop new conceptual knowledge – is ineffective for beginner pupils. In fact, using such inquiry-based approaches for novice learners tends to lead to little, if any, learning (Kirschner et al., 2006). Any knowledge that pupils do acquire is likely to be incomplete, disorganised, and intermixed with misconceptions.

Instead, it is more effective to provide explicit instructional guidance to novice pupils, by a) providing clear explanations and b) modelling the skills you want pupils to acquire, rather than asking them to work these things out for themselves. This is seen in the *worked example effect*: providing pupils with a worked solution (for example, a narrated step-by-step guide to balancing symbol equations) reduces the risk of cognitive overload, as their attention is directed towards the relevant aspects of the process.

To be clear, activities such as independent tasks, problem-solving, pupil collaboration, or practical work are not worthless – far from it – but to be effective such approaches should come *after* a period of initial, explicit instruction, not before. In fact, the *expertise-reversal effect* suggests that as learners gain expertise and their schemas within a domain develop, the utility of explicit instructional guidance decreases, and practising problem solving becomes a more effective learning approach (Sweller et al., 2019).

CONSOLIDATING LEARNING

Whilst the availability of storage space in long-term memory is not something you need to worry about, pupils' ability to recall learnt information *is*. Over time, we appear to forget what we know, a natural process that cannot be avoided.

Yet as a teacher, you want to *consolidate* pupil learning, ensuring that changes to their long-term memories are as long-lasting as possible.

Studies suggest that learners commonly adopt the revision practices of re-reading their notes or highlighting key information. However, these are relatively ineffective at consolidating learning (Dunlosky et al., 2013). Re-encountering materials leads to a *familiarity effect* – the feeling of 'Ah yes, I knew that!' masks the fact that if we'd been asked to recall the information beforehand, we wouldn't have been able to.

Instead, the act of *retrieval practice* is key to strengthening existing knowledge in long-term memory. Studies have compared the repeated studying of materials (i.e. cramming) with replacing one or more of those study sessions with self-testing episodes.

In the short term, cramming can lead to marginally better outcomes in test scores, but re-assessing learners after a delay of a couple of days finds that the 'crammers' have already forgotten much of what they appeared to have learnt and are out-performed by the 'self-testers'. This is the *testing effect*, with the stark drop-off in retention for the cramming condition clear in Figure 9.2.

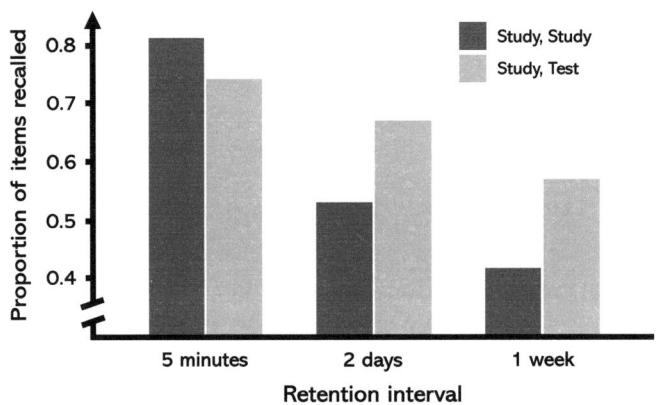

■ **Figure 9.2** A comparison of final test scores at various intervals for participants who conducted two back-to-back study sessions with those who conducted a single study session followed by a self-testing session (Roediger & Karpicke, 2006)

This effect is borne out in both lab-based and classroom studies (Adesope et al., 2017), and the subsequent learning gains mean that leaners both remember knowledge for longer and re-learn seemingly 'forgotten' information more quickly.

Evidence also suggests that the more effortful the act of remembering is, the greater the consolidation of knowledge in long-term memory. The easiest way to affect this is to leave a space in time between initial instruction and the retrieval attempt. This is *distributed practice* and is a simple yet powerful tool: using the power of forgetting to support pupils to consolidate their learning over time (Roediger & Pyc, 2012). Another factor that can affect the difficulty of retrieval is the availability of cues. Key information displayed on a wall, or the use of an open book or knowledge organiser diminish the effort required for recall, thereby decreasing its consolidative power.

Unfortunately, there is no simple or fixed answer to the question of the 'optimum time' to leave before retrieval, as the rate of forgetting depends on factors such as the complexity of the information, pupils' prior knowledge, or the number of previous revisits.

Task 9.4 **Planning for retrieval**

Retrieval practice itself will not benefit pupils if you do not carefully consider *what* is being recalled. For this to be a valuable use of teaching time, choosing the most important content to focus on retrieving is crucial.

1 Reflect: are starter quizzes, 'do now' activities, or low-stakes recall tests embedded as a regular part of the curriculum in your current school?
  i   If so, consider: what choices have been made about the knowledge chosen for recall? What are colleagues and pupils' views about them?
  ii If not, work with your mentor to introduce regular low-stakes quizzing into your teaching practice to consolidation pupils' learning. You should prioritise key knowledge or concepts that are fundamental, commonly needed, or transformational for pupils' abilities to access the science curriculum.

2 Reflect: in your current long-term planning, or the curriculum plans in your school's science department, have opportunities been intentionally designated for pupils to effortfully recall key, previously learnt knowledge at regular intervals throughout the year, rather than viewing each learning episode as a one-off?
  i   If so, consider: what choices have been made about the spacing of these retrieval opportunities over time? You may benefit from discussing this with your mentor or other colleagues.
  ii If not, work with your mentor to introduce spaced retrieval of key pupil learning into your teaching practice over time.

SUMMARY AND KEY POINTS

- Learning is the process of changing the knowledge that is stored in long-term memory. This change is mediated by two important gatekeepers to learning – attention and working memory
- Our minds can only attend to a handful of things at a time. We can process images and verbal information simultaneously, but too many sources of either can quickly lead to overload and prevent learning
- Expertise in a domain is a proxy for extensive, well-connected schema in long-term memory. Without these in place, novice pupils are unlikely to learn from problem-solving or other minimally guided approaches
- A central role of teachers is to break down and incrementally develop the knowledge and skills to build pupils' mental models over time and to build in distributed opportunities to retrieve this knowledge in order to consolidate it.

Check which requirements for your initial teacher education course you have addressed through this chapter.

FURTHER RESOURCES

Didau, D., & Rose, N. (2016). *What every teacher needs to know about psychology*. John Catt.

A highly detailed, comprehensive account of insights from the field of psychology to help classroom teachers better question and develop their teaching practices, covering everything from evolutionary psychology to neuroscience and, of course, plenty of cognitive psychology.

Adam Boxer's blog, https://achemicalorthodoxy.wordpress.com/

The blog of Adam Boxer, a science teacher and leader who shares clear and well-evidenced examples of how he incorporates the principles and implications of cognitive psychology into his teaching.

Howard-Jones et al. Science of Learning resources, https://www.scienceoflearning-ebc.org/services-2/

A range of resources from Paul Howard-Jones and colleagues that consider the implications of findings from neuroscience for the practice of classroom teachers.

Kirschner, P. A., & Hendrick, C. (2020). *How learning happens*. Routledge.

In this rigorous yet accessible book, Paul Kirschner and Carl Hendrick have summarised the key findings from 28 of the most seminal works in educational psychology and unpicked what these mean for your teaching practice.

CONSTRUCTIVIST APPROACHES TO LEARNING AND TEACHING SCIENCE

Richard Brock

INTRODUCTION

The learners in your classroom will have formed ideas about the scientific concepts on the curriculum before you teach them. These ideas come not only from their previous learning in primary schools, but through their everyday experiences with materials and living things. Your pupils will have pushed and pulled objects and become able to predict how they will move, they will have observed the behaviour of animals and developed their own views on how to classify living things, and they will have experienced different kinds of materials and their properties. The ideas learners develop outside of the classroom can have a powerful influence over how they interpret the new concepts they are taught by their teachers. For example, a pupil may come to believe that electricity behaves somewhat like a liquid – people talk of a flow of electricity and wires can seem to act like pipes. This model of electricity will have significant consequences for how the learner interprets their teacher's explanations of electrical circuits. The observation that learners develop their own understandings of the world which can differ from scientific models and that those ideas can have consequences for teaching is a central principle of the constructivist view of learning.

Constructivism has been referred to as a 'grand unifying theory' of science education (Colburn, 2000). This chapter will introduce the assumptions of constructivism as a theory of learning and consider the implications of the theory for practice in the science classroom. Constructivism has been interpreted in a number of ways, both by its supporters and its critics, and a range of different teaching approaches have been described as constructivist techniques. This chapter aims to present an introduction to the consensus principles of constructivism. Different interpretations of constructivism will be discussed and the major criticisms of constructivism addressed. The second half of the chapter considers the implications of constructivism for your practice and suggests

DOI: 10.4324/9781003110187-11

some ways in which constructivist principles can be put into practice in your teaching.

OBJECTIVES

At the end of this chapter, you should be able to:

■ Understand what constructivism is and how it applies to teaching and learning
■ Know the key features of the constructivist models of learning developed by Piaget and Vygotsky
■ Be able to use constructivist ideas to inform the planning of science lessons.

WHAT IS CONSTRUCTIVISM?

Writing in the fourth century BCE, Aristotle compared the human mind to a writing tablet which starts as a blank sheet of wax before marks are inscribed on it with a stylus. Prior to teaching, Aristotle argued, the mind has no content or structure, an idea that came to be known as the blank slate (or tabula rasa) model. Whilst the blank slate model influenced a number of educational thinkers, evidence from cognitive psychological and linguistic research suggests that certain abilities, for example the ability to acquire language (Pinker, 2002), are present from birth and the blank slate model is now rejected by many psychologists and philosophers. In contrast to the blank slate model, research suggests that humans have innate intuitions about the world that lead learners across cultures to make similar assumptions (Shtulman, 2017). In the context of science classrooms, for example, young children typically assume that objects that seemingly move by themselves (for example, flames and clouds) are alive, whilst those that do not (for example, trees) are not alive. Whilst different versions make different claims, constructivist models of learning tend to be based on two assumptions:

■ Learning is process in which pupils seek to make sense of new information in relation to the knowledge they already possess to construct personal understandings
■ Learners develop ideas related to concepts on school curricula before they have experienced formal teaching which have consequences for how they learn academic concepts (Taber, 2009).

When discussing constructivism, it is important to distinguish among three different uses of the term: a philosophical position on knowledge, a psychological description of learning, and a teaching approach. This chapter first examines constructivism as a model of learning and then the implications of that model for teaching practice.

Key constructivist thinkers

Often cited as a founder of constructivism, Jean Piaget was a Swiss psychologist who was interested in children's cognitive development. He noted that, in making sense of the world, young children developed frameworks that organised their ideas, which he labelled schemata (note that schemata is the plural form of the singular schema), now more commonly referred to as conceptual structures. For example, Piaget observed that a child may make sense of their observation that the sun seems to follow them wherever they go by adding the sun to the schema that includes living things (Piaget, 1929). Piaget was interested in how schemata are modified, and he proposed that learners can respond to new information in two ways: by assimilation, when the information is interpreted using an existing schema; or by accommodation, when a schema is modified to fit the new stimulus. For example, when encountering the concept of a whale for the first time, a learner might easily assimilate the new concept into the schema of fish (as whales have fins and live in water). After further learning, they may modify their schema of mammals and reclassify whales as mammals, a process of accommodation. Learning, in Piaget's view, can be imagined as passing through phases of equilibrium and disequilibrium as new information that coheres or contrasts with existing schemata is encountered and processed. The observation that existing schemata influence how new material is processed suggests that teachers develop an understanding of their learners' schemata and plan their teaching to allow new concepts to be readily accepted into already existing knowledge structures.

Whilst Piaget's work focused mainly on individuals, Lev Vygotsky emphasised the role that the relationships between people play in learning. The learning theory he developed therefore became known as social constructivism. Vygotsky (1931/1981) argued that learners often first encounter new concepts in their interactions with other people, for example, in playing with friends or through reading others' ideas in books. He drew a distinction between the spontaneous ideas learners develop without being exposed to formal teaching – which he called spontaneous concepts – and the academic concepts introduced in schooling (Vygotsky, 1962). Learning, in Vygotsky's model, involves an interaction between spontaneous and academic concepts. Pines and West (1986) illustrated this assumption by comparing learning to the growth of two vines, an upward growing shoot from a learner's spontaneous ideas and a downward growing vine from formal academic concepts. Learning, in this model, involves the integration of knowledge gained from these two routes. Vygotsky emphasised the role of more knowledgeable others (for example, parents and teachers) in supporting learners to develop their ideas. Learners can complete some tasks successfully by themselves, and some tasks are impossible even with help. Vygotsky emphasised a group of activities, which he labelled as existing in the zone of proximal development, which a learner can complete when supported by a more knowledgeable other, like a teacher. These tasks, pitched at an intermediate level of challenge, are ideal classroom activities. Vygotsky's zone of proximal development inspired the pedagogic approach of scaffolding (Wood et al., 1976) in which a teacher pre-empts the challenges that a pupil may encounter in learning a concept and plans activities to support the learner to meet those demands.

Task 10.1 **Learning about the constructivists**

This is an academic task that provides theoretical background that is useful for understanding claims for the origin of constructivist pedagogies. An awareness of the assumptions of different thinkers can help you to critique and adapt their ideas for your own practice.

- ▪ Choose one or two of the constructivist thinkers from this list: David Ausubel, Jerome Bruner, John Dewey, Rosalind Driver, George Kelly, Maria Montessori, Jean Piaget, and Lev Vygotsky
- ▪ Research the thinker's contributions to constructivism and write a paragraph describing their key ideas. Second, research how the educationalists ideas have been applied in the science classroom. Where possible, ask an experienced teacher for their interpretation of the thinker and how they have applied the theories in their own practice. Conclude by reading about some of the thinker's critics and summarise their arguments. No educational theory is a perfect representation of learning and it is helpful to consider the strengths and weaknesses of models of learning.

Critiques of constructivism

Constructivism has been a highly influential theory in science education, and a number of criticisms of its assumptions and its application to practice have been raised. First, it has been claimed that some constructivists value learners' ideas too highly in comparison with scientific ideas (Matthews, 1993). Whilst that critique might apply to some models, many teachers and researchers accept that whilst an understanding of learners' conceptions is valuable for supporting teaching, it does not follow that alternative conceptions are of equivalent value to scientific knowledge (Taber, 2009). Second, whilst he acknowledges the usefulness of constructivism as a model of learning, Mayer (2009) has critiqued the effectiveness of some teaching approaches that are claimed to be based on constructivist principles, for example, unguided discovery learning in which a learner is left to independently discover scientific concepts. Evidence suggests unguided discovery learning has limited effectiveness in practice (Kirschner, Sweller, & Clark, 2006), but the claim that the approach is based on constructivist principles misinterprets the assumptions of the theory. The constructivist model of learning suggests that, without guidance, learners are likely to develop understandings that differ from the scientific position. Another example of the misinterpretation of constructivism occurs in pedagogies that assume that the claim that learning is an active process means that effective learning requires classrooms be filled with noise and movement. The observation that the construction of schemata is an active process emphasises that new information is processed and interpreted by learners, leading to novel interpretations that may not have been intended by the teacher. Activity then refers to cognitive processing – rather than physical activity – and can occur whilst a learner is silently listening to a teacher or reading a book. Despite critiques of certain pedagogical interpretations of constructivism, there is a robust body of evidence that supports the two assumptions of constructivism as a model of learning given earlier (learners develop their own understandings of the world and those ideas have consequences for learning; Shtulman, 2017). Even critics of some pedagogic interpretations of constructivism accept constructivism as a theory of learning (Sweller, 2009). Whilst there are a number of plausible suggestions (see later mention), there is, as yet, no clear consensus on effective teaching approaches based on the assumptions of the constructivist learning theory.

Task 10.2 **Science teachers' interpretations of constructivism**

Science teachers can draw on a range of different models of learning to inform their teaching. Currently, learning theories based on cognitive science research are enjoying renewed interest and are shaping approaches to classroom practice (See Chapter 9: Science of Learning). For example, models of teaching based on cognitive load theory, which argues that learners have limited working memory capacity, suggest that activities should be designed so that pupils are not required to hold too many concepts in mind at the same time. Cognitive models of learning are not incompatible with the constructivist model and good classroom practice tends to cohere with both models of learning. To develop your own understanding of learning and to gain practice with applying theory to practice, it is useful to survey the theoretical underpinnings practicing teachers use to guide their teaching. Choose a couple of members of your science department and ask them:

1 What assumptions about learning guide your practice? How do you use theory to guide your practice?
2 To what extent is constructivism a useful model of learning for science teachers?
3 How can constructivist assumptions about learning be used to guide effective teaching practice?

CONSTRUCTIVISM IN SCIENCE EDUCATION

An empirically well-supported finding of constructivist research is that pupils tend to develop ideas about the world that differ from scientific concepts and are resistant to change (Shtulman, 2017). For example, learners often believe that when a ball is swung round on a string, if the string breaks, the ball will follow a curved path, rather than the straight-line trajectory predicted by Newtonian mechanics. Learners may believe that gases don't have mass or that plants obtain all that they need to grow through their roots. Such ideas are referred to as intuitive ideas, misconceptions, or, as in this chapter, alternative conceptions. Researchers have catalogued the alternative conceptions held by pupils of different ages in many topics on the science curriculum (See, for example, *Making Sense of Secondary Science* (Driver et al., 2015) and the American Association for the Advancement of Science (AAAS) website linked to in the Further Resources list at the end of this chapter.). Knowing the typical alternative conceptions pupils hold in a topic can help you plan lessons that support pupils to inhibit those conceptions and preferentially activate scientific concepts. Rather than relying on the patterns reported in existing research, it can be informative to assess the prevalence of alternative conceptions in your classes. A number of resources exist to assess students' alternative conceptions including the diagnostic questions developed by the University of York's Best Evidence Science Teaching (BEST) project and those available from the AAAS (see links at the end of the chapter).

Task 10.3 **Find out about your pupils' alternative conceptions**

Choose an upcoming topic that you are going to teach and use a book such as *Making Sense of Secondary Science* (Driver et al., 2015) or *Scienceblind* (Shtulman, 2017) to research common alternative conceptions of that topic. Then, either use existing diagnostic questions from the Best Evidence Science Teaching (BEST) or American Association for the Advancement of Science (AAAS) websites (see links in the resources section, and Chapter 21 for more information on BEST) to assess what alternative conceptions are held by pupils in your class. You might use open-ended questions to gather detailed information on pupils' views and reflect on any differences between your assessment data and the findings reported in the literature. It can be useful to track changes to the prevalence of these conceptions over a sequence of lessons by giving pupils a repeated assessment at the start and end of a topic.

Applying constructivism in the classroom

Much research suggests that alternative conceptions, for example, the belief that gases have no mass, are highly resistant to change (Shtulman, 2017). Such inertia occurs because the ideas often develop from years of observation and engagement with the world and because the conceptions are, at least partly, tacit, that is, pupils may not be able to explicitly state the conception in words. Therefore, don't feel too disheartened if your pupils revert to using alternative conceptions after a period of using scientific concepts successfully. Even experts scientists, if pushed to answer questions rapidly, relapse to using alternative conceptions that it might be imagined had been entirely eradicated (Goldberg & Thompson-Schill, 2009). Rather than conceptualising learning as the replacement of alternative conceptions with scientific concepts, the process might be better thought of as a gradual increase in the likelihood of use of scientific models and an inhibition of intuitive conceptions. Plan to take alternative conceptions into account by challenging, refining, or elaborating on them to develop expert models (Shtulman, 2017). The next section discusses four approaches, inhibition, bridging analogies, self-explanation, and knowledge organisers, that you can use to support pupils to reduce their use of alternative conceptions and become more likely to use scientific concepts, a process referred to as conceptual change.

Inhibition

Emerging evidence suggests that the ability to inhibit (or suppress the use of) alternative conceptions supports the learning new scientific concepts (Brookman-Byrne, Mareschal, Tolmie, & Dumontheil, 2018). After teaching the target scientific concepts and allowing pupils time to practice their application, conceptual change can be supported by giving pupils the opportunity to practice inhibiting alternative conceptions. One approach is to encourage pupils to become aware of when they use alternative conceptions, to highlight how they differ from scientific ideas, and to practice substituting the scientific concept (Hammer & Elby, 2003). For example, when learning about electrical current, some pupils believe current behaves like a fluid that is used up as it travels round a circuit. You might, after teaching the scientific model of current and giving plenty of opportunity for practicing the accepted model, describe the alternative conception and model the process of activating and inhibiting the conception in an example problem ('In this series circuit, the current will fall at the first bulb, ah, no – that is 'the current is used up' alternative conception – the current will be the same at all points in the circuit'). Then you can set some problems that allow pupils to practice the skill of inhibiting their alternative conceptions. This process can be likened to inoculation – just as patients are exposed to a weakened form of pathogen to prevent illness, pupils can be introduced to alternative conceptions and practice inhibiting them. Note that, given the tenacious nature of alternative conceptions, inhibition is a process that requires practice and effort. It is worth conceptualising your role as first to teach scientific knowledge (that is, of course, an important aim), but also to help pupils to develop their ability to inhibit alternative conceptions.

Bridging analogies

Learners often develop multiple ways of interpreting the world and can simultaneously hold both scientific concepts and alternative conceptions that are selectively activated in different contexts. This can lead to frustration in teachers when pupils appear to have learned successfully, for example they might use the scientific concept to respond to some questions in class, but then, on encountering a new context, for example, an exam question, revert to using an alternative conception. One approach to support pupils to apply scientific concepts to a wider range of contexts is the bridging analogy. A bridging analogy is a progression of teaching activities that encourages a student to apply knowledge that is well understood in one context to another, less well-understood, situation (Clement, 1993). For example, pupils may find it difficult to understand that surfaces can exert an upward force, called the normal reaction force, on objects resting on them. Intuitively, it can be challenging to see how an inert object like a table exerts an upward force and pupils at early stages of their learning about forces can fail to add the normal reaction force to force diagrams of objects resting on surfaces. Clement (1993) suggests that teaching to support the understanding of the normal reaction force might begin from a context in which pupils are more likely to activate the scientifically correct understanding of force – a pupil given a compressible spring is asked to push down on it and describe what they experience. The pupil is likely to report that the spring exerts an upward force on their hand. Clement (1993) labels this idea an anchoring intuition, an activation of the target scientific concept in one context. This experience may, by itself, be insufficient to allow learners to correctly identify the forces that act in other related contexts, for example, when a book rests on a table. Whilst experts might be able to appreciate their similarities, for novice physicists, a spring and a table are significantly different objects. To support pupils to transfer their anchoring intuition beyond the context of the spring, the learner can be shown first a book resting on a number of compressible springs (like a spring mattress), a book resting on some foam, a book on bendy piece of plywood and finally a book on table. The sequence of situations is designed so that each context is somewhat similar to the last in order to encourage the transfer of the concept of the normal reaction force from the context of the spring to the case of a book on the table. Similar sequences of bridging analogies can be designed for other contexts. For example, pupils may develop the alternative conception that material disappears when it dissolves. An anchoring example in this context is the dissolution of potassium permanganate in water – the colour change can act as a cue to inhibit the alternative conception that the substance has disappeared. A useful bridging context is the dissolution of unrefined sugar in water, which leads to a noticeable colour change but is not too distinct from the case of dissolving white sugar in water.

Self-explanation

Constructivists argue that pupils process new information to develop personal understandings of academic concepts. It is plausible, therefore, that giving pupils time to engage in sensemaking, by developing their own interpretations of the new information and ensuring it coheres with their existing schemata, followed by appropriate feedback, might support learning. Indeed, a body of research supports the effectiveness of self-explanation as a teaching approach (Fiorella & Mayer, 2016). Self-explanation can help pupils to structure new material and to spot alternative conceptions or inconsistencies in their own arguments. Whilst some learners may self-explain naturally, for example, by asking themselves 'does this make sense?' when reading a text, many pupils will need external prompts to encourage the behaviour. In the classroom, a self-explanation activity might be structured as follows: first you explain a new idea, for example, you might introduce the concept of natural selection; second, you ask your pupils to produce, orally or in writing, an explanation of the concept in their own words, and finally, you give feedback on those explanations, correcting any misunderstandings and improving the quality of the argument. The approach can also be applied to problem solving – when teaching how to solve a new kind of problem you might ask pupils to explain the decisions they are making. You can model the process of self-explanation by 'thinking aloud' when you are problem solving or showing the kind of thought processes you would go through when encountering a new scientific concept. Self-explanation is more likely to be effective when it is a routine expectation, rather than a strategy that is used occasionally. It is related to the idea of metacognition and self-regulation (see Chapter 1).

Knowledge organisers

Whilst the strategies introduced so far have focused on individual concepts, since Piaget's work on schemata, constructivists have conceptualised learning as the development of organisations of concepts, more commonly now referred to as conceptual or cognitive structures. Expertise in scientific domains arises not just from knowing scientific concepts but also in understanding how those concepts are related. You might usefully conceptualise teaching as having two phases that somewhat overlap: first pupils are taught some new concepts in a domain and second they are supported to relate the new concepts to existing knowledge and to apply the concepts in an expert-like manner (Brock, 2018). The development of conceptual structures is significant because the structure of prior learning can guide the way new information is processed. Piaget (1977) argued new knowledge is more readily accepted by learners when it resembles, to some extent, what they already know. This claim was developed into a teaching approach by David Ausubel (1960) who proposed the idea of advance organisers, information introduced by a teacher that helps pupils to categorise and retain novel concepts. For example, before teaching about the categorisation of whales as mammals, you might point out that whales give birth to live young and nurse their calves with milk. Information that emphasises whales' mammalian characteristics can help pupils to appropriately categorise whales in their conceptual structure and helps to inhibit the alternative conception that whales are fish. Alternatively, in the context of teaching about light, a simulation that models the behaviour of balls when they bounce off a surface might be used as an advance organiser for teaching the principles of reflection.

Pupils' appreciation of conceptual structure can be supported through a visual representation, concept maps (Edmondson, 2005), diagrams in which concepts are linked together with lines labelled with their relationships (see Figure 10.1).

Concept maps can be used to support pupils to develop expert-like conceptual structures in a number of ways including by showing pupils pre-drawn concept maps which represent expert knowledge structure or as assessment tools by asking pupils to represent the relationships between concepts. Pupils might compare the concept maps they produce with those of experts so that they get a sense of where their conceptual structure requires revision.

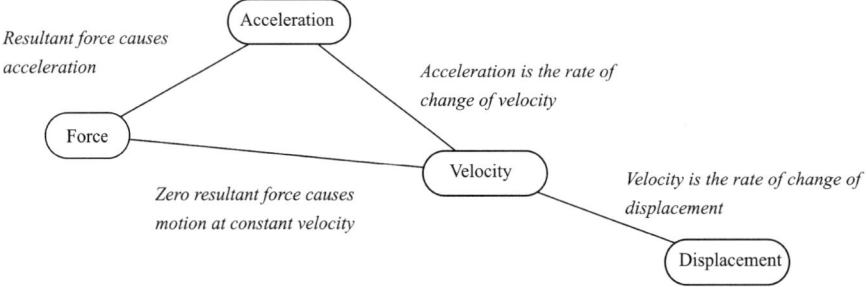

RICHARD BROCK

■ **Figure 10.1** An example concept map in the context of teaching about force

---

Task 10.4 **Observing and/or planning a science lesson using constructivist principles**

First, observe a science lesson and consider what constructivist principles have been drawn on in the lesson plan and execution. You might consider whether the following points have been addressed:

■ How has the teacher taken pupils' prior knowledge into account?
■ How is scaffolding used to ensure tasks are pitched at an appropriate level of challenge?
■ What strategies has the teacher used to support the development of conceptual structures?
■ Can you observe any instances in which students' alternative conceptions have impacted how they learnt new material?
■ What strategies has the teacher used to prompt conceptual change?
    Choose an upcoming lesson and consider how you can use constructivist principles to guide your planning of activities.
■ How will you assess or research pupils' current understanding of the topic and uncover any alternative conceptions about the domain they hold?
■ How might any alternative conceptions impact on pupils' ability to accept scientific concepts?
■ How can you plan for activities that allow pupils the opportunity to practice inhibiting their alternative conceptions?
■ How can you prepare pupils' conceptual structures to readily accept the concepts you will teach?
■ How can you support pupils to organise their conceptual structure in the same way as experts?

SUMMARY AND KEY POINTS

■ Constructivism is a model of learning that assumes that learners develop understandings of scientific concepts before they encounter formal science teaching and that these conceptions may differ from accepted scientific models.

- Pupils' alternative conceptions can be very difficult to change and even experts may, on occasion, revert to using alternative conceptions
- You can guide your teaching by researching common alternative conceptions in a topic or assessing your pupils before teaching to uncover the prevalence of different conceptions
- Emerging evidence suggests pupils can be supported to transition from the use of alternative conceptions to scientific concepts by being given the opportunity to practice noticing when they activate alternative conceptions and substituting a scientific alternative
- Plan for teaching approaches that help pupils organise the knowledge they have been taught.

Check which requirements for your initial teacher education you have addressed through this chapter.

FURTHER RESOURCES

American Association for the Advancement of Science (AAAS) – Science Assessment. (http://assess.bscs.org/science/pages/about) The AAAS pages give a list of topics which you can use to explore common alternative conceptions and their reported prevalence in different age groups. The website also allows you to make custom multiple-choice assessments to diagnose alternative conceptions and compare your students' performance against their peers.

Best Evidence Science Teaching (BEST) – University of York (https://www.stem.org.uk/best-evidence-science-teaching). A series of research-informed resources, including recommendations for a sequence in which to teach concepts, diagnostic tools to probe pupils' alternative conceptions, and activities to promote conceptual change. The resources are targeted at age 11–14 science.

Driver, R. et al. (2014). *Making sense of secondary science: Research into children's ideas.* Routledge. This book catalogues a large body of research into pupils' alternative conceptions across a range of topics on the primary and secondary science curricula.

Mintzes, J., Wandersee, J., & Novak, J. (1998). *Teaching science for understanding: A human constructivsit view.* Academic Press. Mintzes and colleagues' book provides a good introduction to constructivist theory but also a number of practical strategies of teaching and assessing scientific understanding. Chapters include focus on a range of topics including: metacognition, using analogies, and knowledge organisers.

Shtulman, A. (2017). *Scienceblind. Why our intuitive theories about the world are so often wrong.* Basic Books. Recommended reading for useful insights for science teachers.

Taber, K. S. (2011). Constructivism as educational theory: Contingency in learning, and optimally guided instruction. In J. Hassaskhah (Ed.), *Educational theory* (pp. 36–91). Nova. Recommended reading to understand the implications of constructivism for learning and teaching.

MODELS AND MODELLING IN SCIENCE TEACHING

Lindsay Hetherington and
Sam Mead

INTRODUCTION

Models are as fundamental to science education as they are to science itself. As we'll discuss later in the chapter, models are 'thinking tools' (Taber, 2017, p. 263), helping scientists to reason about concepts and phenomena in the world around us. This is very similar to how models are used in the classroom: teachers introduce pupils to models that will help them to think about and understand the material that makes up the curriculum. As Willingham (2009, p. 18) says, 'memories are formed as the residue of thought', meaning that when science teachers support pupils to think hard about a scientific idea using models, they are supporting their learning of scientific content as well as their understanding of scientific processes. Models in science are not simply representations of a phenomena; they are *explanatory* and *exploratory*: they take complex systems, break them down into understandable components, and enable scientists to investigate a phenomenon. The same applies to the use of models in science education, where they can be used to frame an explanation and to support pupils' investigation.

An EEF review of current evidence from educational research identified the use of models to support understanding as a key strand in effective science teaching, arguing that science teachers need

> to use models to help pupils develop a deeper understanding of scientific concepts; that they need to select the models they use with care, and that they should explicitly teach pupils about models and encourage pupils to critique them.
>
> (Education Endowment Foundation, 2018, p. 8)

This final point is crucial. Using models as part of an explanation is very helpful, but teachers must be careful to avoid reifying the model so that pupils

    DOI: 10.4324/9781003110187-12

believe the model *is* the phenomenon itself and is a 'true' representation of the world. For example, all too often, pupils can be disillusioned when they meet different, more complex models as their school career progresses. They are often put out that we, as teachers, have 'lied' to them about the truth of the world. This undesirable situation arises when we aren't explicit enough when teaching the nature of models to our pupils, running the risk of frustrating them by appearing to teach disposable ideas.

> After 10 years of working with new science teachers, I am still struck by how many of them portray their scientific careers as a series of stages in which they found out that everything they had been told previously was wrong.
>
> (Hardman, 2017, p. 91)

It's our responsibility as science teachers to take the opportunity to teach about models, an important and often overlooked aspect of the nature of science. A classic example of this is the developing model of atomic structure used in chemistry (see Chapter 6 for a more detailed discussion). Teaching about the nature of models shows pupils that the basic models they have been working with in the early part of their science learning aren't wrong; they're simply the most appropriate model for the given use case – pupils don't need a full electronic configuration model of the atom to explain the essential behaviour of solids, liquids, and gases, for example. It's important for science teachers to make this explicit.

When we teach our pupils the nature of models, we allow them the opportunity to understand and evaluate them. This means that alongside models-based teaching, where teachers and pupils use models to help explain and understand concepts, teachers employ modelling-based teaching strategies, where pupils create their own models in the classroom and then use those models to augment their learning (Gilbert & Justi, 2016). Modelling-based teaching draws on constructivist learning theory in a manner that enables pupils to work metacognitively with the models they are using to explain and understand.

In this chapter, we will explore models-based and modelling-based science teaching. We will begin by discussing what a scientific model is and how this relates to the kinds of models we teach *about* and teach *with*. We then explore how models can be used in the classroom.

OBJECTIVES

At the end of this chapter, you should be able:

■ To explore the nature of models in science and how that relates to models used in science teaching
■ To consider how the use of models relates to key theoretical ideas about learning, particularly constructivism
■ To discuss the relationship between models and modelling, the nature of science, and the nature of learning

WHAT IS A MODEL?

Models as representation + affordance

The definition of the word 'model' is highly context dependent. As noted by Gilbert and Justi (2016, p. vii), there are three clear tiers of definitions for the word model:

- Those that occur only in everyday life, for example, someone who wears clothes to display them
- Those that are relevant both in everyday life and in science, for example, the layout of a particular passenger aircraft
- Those that exist in everyday life but have special status in scientific contexts, for example, the Watson-Crick-Franklin model of DNA.

Within the context of science education, we can most simply define models as *representations* of scientific concepts. By this definition, models represent physical phenomena in alignment with scientific consensus. However, models have utility beyond simply representation, as they are used in science to develop and test explanations. The need for models often arises due to the fact that scientific concepts are abstract and cannot be easily observed: models are often simpler than the real-world phenomenon or enable us to think about phenomena occurring at scales that are not easily observable. Johnstone's triangle, or the Chemistry Triplet (Johnstone, 1982, cited in Taber, 2013; Figure 11.1) is a useful way of looking at this. At the apex of the triangle is the macroscopic phenomena

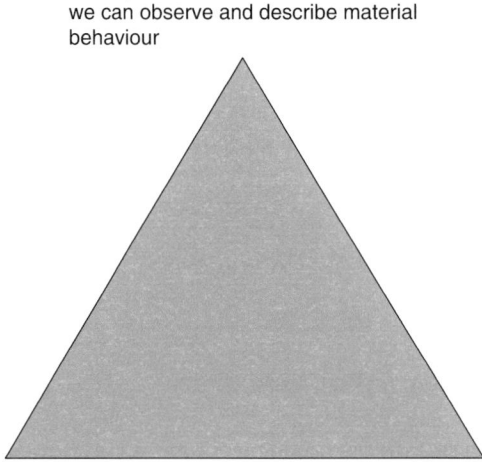

Descriptive/Functional level at which we can observe and describe material behaviour

Explanatory/Model level at which we attempt to explain the observed phenomenon

Representational level at which we try to represent behaviour using symbolic formulae and equations

▪ **Figure 11.1** The position of modelling within Johnstone's triangle/the chemistry triplet, after Johnstone (1982) cited in Taber (2013)

we might be able to observe in a classroom experiment, such as the change in properties of iron and sulfur when reacted to produce iron sulfide – the descriptive or functional level. At the bottom left we have a representation of the reaction using a particle model – the explanatory level. At the bottom right is the symbolic representation of the reaction that we use in chemistry – the representational level. Both the bottom corners of the triangle are representations of the reaction, but the particle model goes beyond representation to explain what is happening and make predictions for the reaction under different conditions or for different reactions.

Scientific models can be *observatory, predictive, or both*. Observatory models, such as the model of the atom, are based on scientists' observations of the world and serve as direct explanations of the phenomena they describe. Predictive models, on the other hand, have led to hugely important discoveries in all disciplines of science. Perhaps the most famous predictive model of all, Mendeleev's Periodic Table, allowed chemists to model the properties of elements before they had been discovered. More recently, predictive models have become increasingly dominant in progressing our understanding of the universe, in many cases determining the outcome of funding applications. The $13 billion spent on the discovery of the Higgs boson is entirely thanks to predictive models, models that explain things before we've directly detected any evidence (Hardman, 2017). Science applies both categories of models on a scale of realism: at one end of the scale, models can be a very close approximation of the phenomena they describe. For example, the way we introduce pupils to the model of the structure and function of the human heart is a very close approximation of reality. When pupils are presented with the task of dissecting a heart, the models prove themselves an accurate guide to reality. At the other end of the realism scale, models are used for a predetermined purpose – they are a useful way of thinking and reasoning about certain aspects of reality. To cite another school-level example, in mechanics, teachers frequently model objects with complex shapes as point masses, and we concern ourselves with forces applied to a centre of mass, ignoring the form of the object. But this doesn't mean that more accurate, more detailed models are closer to reality. Ultimately, all models are reductions, so while it's important for the model to resemble the phenomenon, the level of detail does not always determine how useful a model can be. This is especially true in making the transition from models in science to models in science education.

> We might say that a photograph gives us a more accurate image of a landscape than a drawing, but a schematic map is likely to be of more use for orienteering.
>
> (Hardman, 2017, p. 92)

We have seen, then, that models are representations of a phenomenon that we can use to help us explain, understand, and predict: The field of modelling in science is wide and complex. In the context of observatory and predicative models, some realistic and others less-realistic, it's reasonable to ask: does science lead models, or do models lead science? The scientific community accepts new models based on how well they portray the phenomenon in question, but this doesn't mean that they are 'right' or 'true' or even that they are the final version of an explanation. The nature of empirical science leaves room for the continual evolution of models. As techniques and technology evolve, our capacity to look deeper and closer at physical phenomena increases. Naturally this leads to the evolution of accepted models, too, as scientists are able to reason about phenomena using more data and detail. The iterative nature of models is an important aspect of both models in science and in science education. This aspect of the nature of science is one area where the national curriculum in the England, for example, does place a focus on the anatomy of models. GCSE pupils are taught (and examined) on how the model of the structure of the atom evolved over time, from Dalton's primitive model to Bohr's model of atomic orbitals with specific energy levels. It's important for teachers to connect this concrete example to the wider practice of evolving models in science.

Ultimately, models become accepted by the scientific community not because they provide the 'real' explanation of a phenomenon but rather because they help us progress our understanding of the universe. We evaluate models based on many criteria: how well they fit existing understanding; how useful they are for their purpose; how far they can be justified to others; how elegant they are; who has proposed the model. Models aren't often developed to describe the world as it is, rather, they help explain the focus of a scientist's attention.

Task 11.1 **Scientific Models**

Without looking anything up, list five scientific models. Choose one that is typically found in your school science curriculum and describe what it represents, what it can explain and why, and whether it is predictive. You could use Johnstone's triangle or the 'chemistry triplet' (See Figure 11.1) to think about the phenomenon and the model at a descriptive, explanatory, and symbolic level.

Models as Metaphor

Models quite quickly become the most natural way of expressing science. For example, to describe atomic structure without relying on language that uses representation is difficult: if we describe an atom as a nucleus of neutrons and protons, surrounded by electrons occupying atomic orbitals, it remains an abstract concept. This leads us to use concrete language to communicate concepts to pupils: our atom becomes a *sphere*, with a nucleus in the *centre*, surrounded by a *cloud* of electrons in different energy *levels*. Models as representations of what a real phenomenon is *like* without actually *being* the phenomenon means that models are rather like metaphors. Indeed, scientists have developed fully fledged scientific models out of an initial use of metaphor or simile to help them visualise what is happening (e.g. Faraday's 'field lines' to think about magnetism). It is interesting to think about the similarities in scientists' development of models to the notion of mental models constructed by pupils discussed in Chapter 10. In the same way that metaphors and similes only work when the reader understands the object on which the analogy is built, the models we are using to teach scientific phenomena will only work if the pupil is able to visualise the model.

By focusing on the language of models, we can compare science education's use of models to English education's use metaphor and simile. Crucially, metaphors and similes are explicitly taught to pupils in English, which allows their *usage* in the classroom without the assumption that they represent the truth. In the English classroom, pupils know that the moon isn't really a wheel of cheese; this is in contrast to the science lab, where many pupils believe that the models they are presented with represent the absolute truth of the world. Models can never be perfect representations of phenomena they describe. Scientists use models for a wide variety of uses and develop them under a wide variety of motivations. Yet the common pupil view of science education suggests that teachers slowly tell pupils the 'truth' about science as they use increasingly sophisticated models. However, this view is a caricature of the reality of science education, as well as of science communication in general (Hardman, 2017). It may help our pupils for us to frame models as metaphorical representations to avoid this misconception of unfolding truth; models are explanations that represent phenomena, but, importantly, they are not the phenomena themselves.

So, if the purpose of models isn't to tell us the truth, how can we use them best to help pupils learn? How can we help them understand models to be a tool used by science to explain certain aspects of the world around us? How can we help them learn to critique and evaluate models?

MODEL-BASED AND MODELLING-BASED TEACHING

Model-based teaching: explaining science using models

Model-based teaching is a standard approach to science education, in which models are used to explain ideas and concepts in science. Models used as part of explanations can be variable, but we can separate them into two types: *scientific models*, which are directly linked either to historic or current science; and models that have been developed specifically for teaching. In the latter category, Taber (2017) distinguishes between *curriculum models* which simplify scientific knowledge in order to minimise the learning demand, or the gap between pupils' current knowledge and current scientific knowledge on a topic, and *pedagogic models (aka teaching models)* which are designed to help learners 'understand the essence of some scientific idea or principle' (Taber, 2017, p. 272). Chapters 5, 6, and 7 in this book all describe some common curriculum and teaching models used in many science curricula. The selection of appropriate curriculum and pedagogic models is crucial in teaching science effectively. As discussed already in this chapter, the use of models in science education reflects an important aspect of the nature of science and it is important that we teach about scientific models and how to evaluate them in order to help pupils understand not only scientific content but also scientific processes. However, choosing models that oversimplify the science risks contributing to the development of misconceptions (see Chapter 10).

Curriculum models therefore need to be true to the heart of the scientific concept being taught in order to provide a basis for future learning that is gradually closer to the current scientific models. The gradual progression of the particle model and model of the atom discussed earlier and in Chapter 6 are good examples of a developing curriculum model that should be articulated to pupils as such.

Task 11.2 **Evaluating a curriculum model.**

Choose a curriculum model that is particularly important in teaching the science subject in which you are most confident. Use the Focus, Action, Reflect (FAR) approach (Walsh, 2019, see the further resources section for more information) to reflect on how and why you are using this model to teach a concept and how well you made use of it in a lesson. The FAR approach asks you to plan, act and reflect in the same way you would about other aspects of teaching, but paying close attention to the use of models. Firstly, *FOCUS* during planning on the concept to be taught using a model, what the pupils already know about the concept or process the model with be describing, and the model you intend to use. Secondly, take *ACTION* during the lesson, discussing with pupils the features of the concept and the model and the similarities and differences between them. Finally, *REFLECT* on whether the model was clear or confusing for the pupils, how it could be improved and whether the class need to revisit the concept.

Teaching models may be physical models, computer simulations, mathematical models, and even virtual reality simulations. For example, the bell jar model of lung action, a model body showing the position of different organs, the rope model used to demonstrate electric current around a circuit, or a computer simulations showing the impact of changing particular reaction conditions on reaction rate are all examples of teaching models. We can categorise these classroom-focused models into different types, which may help us consider the best way to use them in our explanations and what to watch out for as we use them to avoid confusion (Table 11.1), of course bearing in mind that any particular teaching model might involve a combination of these.

▪ **Table 11.1** Types of teaching models

Type	Description	Example/s
Scale models	Used to help pupils visualise relative sizes or timescales – the very small or very large.	Scale of the solar system, the number of particles in a mole, relative sizes of subatomic particles v. the atom, geological time.
Metaphorical models	Also known as analogue models, helping pupils think about something using a more familiar or concrete analogy.	The flow of electricity using the rope model or water model, the structure of the earth using eggs or apples.

(Continued)

■ **Table 11.1** (Continued)

Type	Description	Example/s
Physical models	Created and demonstrated by the teacher or by the pupils themselves, physical models are 3D analogues. Can be static or dynamic, with the latter enabling a process to be shown.	Party popper model of earthquake prediction; anatomical models; model cells; model atoms; bell jar model of lungs; pupil modelling of particles and circuits; molymod models or chemical structures.
Computer simulations/ virtual laboratories	Uses a computer model to enable pupils to explore and investigate by manipulating variables without conducting a laboratory activity. Used where laboratory activities are dangerous or perhaps unreliable or equipment is limited.	Simulations of effect of changing forces on motion, evolution of a system (e.g. peppered moth), chemical reactions such as titrations.
Mathematical/ predictive models	Using a systematic/ mathematical model to demonstrate a concept.	Using dice to model radioactive decay and half-life.
Historical models	Models that were previously proposed as scientific models that have since been superseded by models that explain more data. Curriculum models often build on or include historical models.	Various models of the atom (e.g. plum pudding model).

Task 11.3 **Reflecting on an example of model use in the classroom**

Select one of the types of models from the list in Table 11.1 and choose a relevant example of it from your own classroom experience as either a teacher or observer. Write a brief account using the following headings:

■ Describe what happened; What learning emerged, for whom, and how do you know?; and

■ How did the model relate to that learning? Use your observations to script how you intend to use a similar model in a future lesson.

As Table 11.1 shows, the need for teaching models might arise for a range of reasons, often associated with simplifying something complex or abstract. One common reason for using teaching models is as a result of the scale of the idea being challenging to grasp. For example, a 'toilet roll of time' model is used to demonstrate to pupils the very short timescale of anthropogenic climate change in comparison to existence of the Earth and life on Earth on a geological or 'deep' timescale (Earthlearningidea, n.d.). Another good example is modelling the arrangement of our solar system. Figure 11.2 shows a common model used to illustrate the scale of the planets in our solar system, relative to each other.

■ **Figure 11.2** An illustration showing the approximate sizes of the planets relative to each other (downloaded from NASA resources, https://science.nasa.gov/resource/solar-system-sizes/)

Outward from the Sun, the planets are Mercury, Venus, Earth, Mars, Jupiter, Saturn, Uranus, and Neptune, followed by the dwarf planet Pluto (NASA, 2003).

The model is simple but useful. It clearly shows pupils the order of the planets and their relative sizes. It also gives an opportunity to teach pupils about the limitations of models – to model the distance between the planets, we would need a significantly wider page. With the Earth the size of a large marble, the Somerset Space Walk models the distances between the planets to scale over a 22-kilometre walk, from Taunton to Bridgewater! Teaching models must be selected and used with real care and attention to avoid causing misconceptions. Being explicit about their relative strengths and weaknesses in what they are able to represent, encouraging pupils to evaluate these strengths and weaknesses themselves, and using a range of different teaching models to access different parts of a target concept can all help minimise the risk and capitalise on the learning that ensues through the use of a teaching model. It is helpful where science curricula explicitly include a stipulation that pupils should be able to understand and explain the limitations of models and modelling (see for example the AQA 2019 Combined Science GCSE specification in England, which lists the limitations of the particle model in physics and the various bonding models in chemistry as part of its examinable content). However, we would argue that such explicit discussion should be included when using teaching and curriculum models regardless of whether it is a formal part of the curriculum. Giving pupils opportunities to evaluate the limitations, shortcomings, drawbacks, and simplifications of the common models we use in the classroom is a further step towards reinforcing the understanding that scientific models themselves are representations of phenomena, each with their own pros, cons, and specific use cases. This moves us away from simply models-based teaching, in which models are used to aid explanation of scientific concepts and towards what Gilbert and Justi (2016) call 'modelling-based teaching', which we will discuss in more detail in the next section.

Modelling-based teaching

A large part of using models when teaching science is the concrete, everyday experiences that underpin them. This use of concrete examples to illustrate and clarify explanations is considered a best practice approach to teaching by researchers in cognitive science (see Chapter 9), yet it is so widespread in its usage that it's rarely discussed and not considered as a formal technique by many educators (Weinstein et al., 2018). The fact that concrete examples – and, by extension, models – are so widespread in teaching can lead to an unintended disconnect when analysing models in learning. Throughout this chapter, we have regularly reiterated the importance of pupils' analysing and evaluating models as they learn science, and in this section we will further expand on this and unpick how we might help pupils do this. This idea is not new. Over 20 years ago, academics lamented that, in many teaching settings, too much focus was placed on the content of the model, rather than the concept of the model (Van Driel & Verloop, 1999). More recently, fortunately for the learning outcomes of our pupils, there has been a considerable shift in the use of models in the classroom, away from the models as content consideration in a straightforward model-based teaching approach and towards a models as 'artefacts' view (Gilbert & Justi, 2016) in which learners test, critique, and evaluate models. Gilbert and Justi refer to this approach as 'modelling-based teaching'. To avoid pupils viewing models as the 'truth' and becoming disheartened when a new one is introduced, it is imperative for science teachers to teach the nature of models and modelling, as well as give pupils the opportunity to *use* the models, not just experience the models. This gives pupils the opportunity to learn to differentiate models from the phenomena the model represents. For example, Rivet and Kastens (2012) point out that models are particularly important for teaching about Earth and space, where pupil interaction with the actual phenomena in question is impossible. They note, however, that 'it is commonly perceived that the model 'tells' the concept, rather than being viewed as a tool for supporting the development of understanding' (Rivet & Kastens, 2012, p. 715). This sense of a simplistic one-way communication from model to learner is problematic in pedagogic terms, as pupils require substantial support to enable them to 'see' the correspondence between entities and to manipulate the model in a way that guides their reasoning process. It also does not reflect how models are used by practicing scientists as a mode of inquiry. Watson and Crick famously developed their double helix structure of DNA as a physical model and a theoretical model simultaneously and in dialogue with other scientists (Watson, 1968). In a similar fashion it has been suggested that model-based learning is most effective when learners have the opportunity to test and critique their own models and those of others through collaborative group work (Coll, France, & Taylor, 2005). How, then, might we build opportunities for pupils to use, critique, and evaluate models into our lessons? One way of doing this is by building opportunities for them to develop, articulate, critique, and compare their own 'mental models' (see Chapter 10) with the scientific models, curriculum models, and teaching models that are being explored in the classroom. Concept cartoons (Naylor & Keogh, 2000) are a useful way of helping pupils engage in dialogue to articulate their ideas based in their own mental models and compare them with the curriculum and scientific models about which they are learning (for more on this, see Chapter 21). A more time intensive way of doing this would be to enable pupils to explore a phenomenon and develop their own model to explain it, then to ask them to critique this model for the observations it does and does not explain. Following this, one would then teach the curriculum model and ask pupils how the new model is different and what observations it can explain in more depth. For example, using a series of activities that can be explained using the particle model (a metal ball and ring, dissolving a coloured solid such as copper sulphate in water, evaporation of acetone on glass evaporating dish, etc.) lends itself to this kind of approach. Clearly it takes time to do this well, and teachers need to make choices about when and where they will commit time to pupils' development and critique of their models in this way alongside other approaches (e.g. allowing time for retrieval practices, modelling of answers, and so on, which is a very different type of teacher modelling than that discussed here and is explained in more depth in Chapter 9).

Task 11.4 **Planning for modelling-based teaching**

Create a medium-term plan for a sequence of three to six lessons on a chosen topic in your specialist science subject. In your plan, you should draw on ideas from other chapters around the implications of the science of learning, adaptive teaching, and assessment, for example, but also build in a planned opportunity for pupils to create, use, and evaluate their own model and the curriculum model you are using to teach the key concepts in the topic. You should use any format for your planning according to your specific ITE programme or context.

Model- and modelling-based teaching requires teachers not only to consider the relationship among the curriculum or teaching model, the scientific model, and the scientific concept being taught but also to ensure that the model is effective in connecting to pupils' own lives and experience (and therefore takes into account pupils' own 'mental models'). Godec et al. (2017) define science capital as 'encapsulating all the science-related knowledge, attitudes, experiences and social contacts that an individual may have' (p. 5), pointing out that using the concept of science capital is a way of promoting social justice for the young people in our science classrooms. Pupils come to the science classroom with an enormous range of science capital. As we have outlined in this chapter, we use models in order to make abstract or complex concepts accessible, for example through metaphor and analogy with something more concrete or familiar. Considering this when designing curriculum and teaching models, bearing in mind that there is often a huge gulf between the science capital of a science teacher and a pupil, ensures that we use models that are accessible to all of our pupils. In reflecting on the social and science capital of our pupils, it's clear that the language of models is one of science's most valuable teaching and learning tools (Harrison & Treagust, 2000) but only when appropriately deployed.

SUMMARY AND KEY POINTS

In this chapter, we have considered what constitutes a model and how it is used in science itself. We have explored the way we use models to teach particular concepts in science but also argued for modelling-based teaching in which pupils are encouraged to critique and evaluate models and understand them as *representations* with particular affordances rather than a 'truth'. To effectively teach using models, we need to explicitly teach pupils what a model is. We also need pupils to know that:

- Models aren't the phenomenon; they're a representation
- Models aren't perfect; they have limitations and deficiencies
- Models can be adapted, iterated upon, and evolved
- It's okay to use a limited model, as long as we understand and acknowledge the limitations.

FURTHER RESOURCES

Justi, R., & Gilbert, J. K. (2016). *Modelling-based teaching in science education*. Springer.

   A key volume that provides an excellent background to the principles underpinning modelling-based teaching and the research on which it is based.

Hardman, M. A. (2017). Models, matter and truth in doing and learning science. *School Science Review, 98*(365), 91–98.

   Written for practitioners, this accessible article outlines the crucial role of models in teaching science and teaching about the nature of science.

Taber, K. S. (2017). Models and modelling in science and science education. In K. S. Taber & B. Akpan (Eds.), *Science education* (pp. 263-278). Leiden.

This chapter offers a clear explanation and summary of how we use models and modelling both in science itself and in teaching and learning science.

University of Colorado Boulder. (2021). *PhET simulations: Interactive simulations for science and math*. https://phet.colorado.edu

These excellent resources provide interactive models and simulations for a range of topics in both science and maths education.

USING DIGITAL TECHNOLOGIES TO SUPPORT LEARNING AND TEACHING SCIENCE

Richard Osborne

INTRODUCTION

One of the biggest lessons learnt from the recent global pandemic was the critical role digital technologies can play in education. At first, digital tools were seen just as a way of connecting those who could not physically meet, but soon they became spaces for deep conversations between the disconnected, new places for personal reflection and development, and creative hubs for producing innovative digital content. For learning and teaching science in particular, the pandemic helped to show what was best about digital technologies in education; from using smartphones to capture science as it happened during daily exercise, to understanding COVID-19 statistics through data patterns and interactive graphs, through to dispelling misconceptions through visualisations of the structure of viruses.

This chapter begins by exploring the nature of digital technologies. After a brief look at some key theoretical perspectives used to understand the role digital technologies play in education generally, it moves on to look at barriers to use that student teachers will encounter, including online safety concerns and the impact of school policies and practices. The chapter provides examples of how digital technologies can enhance aspects of teaching, such as assessment, collaboration, investigation, and motivation. Finally, it briefly explores potential future opportunities for learning with digital technologies.

OBJECTIVES

At the end of this chapter you should be able to:

- Consider the nature of digital technologies and what makes them different from their real-world equivalents
- Apply theoretical approaches to your integration of digital technologies in the science classroom

DOI: 10.4324/9781003110187-13

- Describe common barriers to successful use of digital technologies within the secondary school context
- Explore key pedagogical dimensions of teaching and learning with digital technologies and how they apply within science education. `

THE NATURE OF DIGITAL TECHNOLOGIES

What digital technologies are may seem obvious – they are computers, smartphones, websites, apps, etc., but those working within the field are not sure that such a simple understanding of digital technologies is useful, and this in turn impacts the ongoing problem of integrating digital technologies effectively into teaching and learning. As Oliver puts it, there exists an ongoing 'failure to explain technology theoretically' (Oliver, 2013, p. 1).

Taking an ecological perspective

Attempts to explain technology theoretically are often aligned to technology and/or pedagogy (Hew et al., 2019). However, rather than consider technology from just one perspective this chapter takes an ecological approach. Hammond (2020) has argued that the ecological approach can provide a powerful framework to understand how digital technologies are used effectively as it recognises the complexity of the environment within which education happens. Teachers are heavily influenced by the environment around them: their peers' understanding of digital technologies, their pupils' knowledge and expertise, the policies and practices of their own institutions, the wider institutional frameworks that govern and direct schools, and the nature of digital artefacts themselves. All of these have an impact on how digital technologies can be used in science teaching.

The chapter draws on the work of Bronfenbrenner whose overall framework, pictured in Figure 12.1 tailored for secondary education, will be used throughout to highlight each part of the ecology, from the broader picture of digital technology use and narrowing down through to the experience of an individual teacher in the classroom.

New opportunities for experience

It is important to recognise the physicality of digital devices as this impacts how they are used; mobile phones and iPads can be portable, whereas interactive whiteboards are so large that they can be used to capture the attention of an entire room, if positioned and used effectively. However, our primary interactions are not with the hardware but the digital places that they create and display on their screens. When we turn on our devices we can click buttons, move sliders, enter text, and suddenly be moving through all types of content without ever leaving our physical place. We experience digital places. The power of digital technologies from a teaching and learning perspective is in these digital

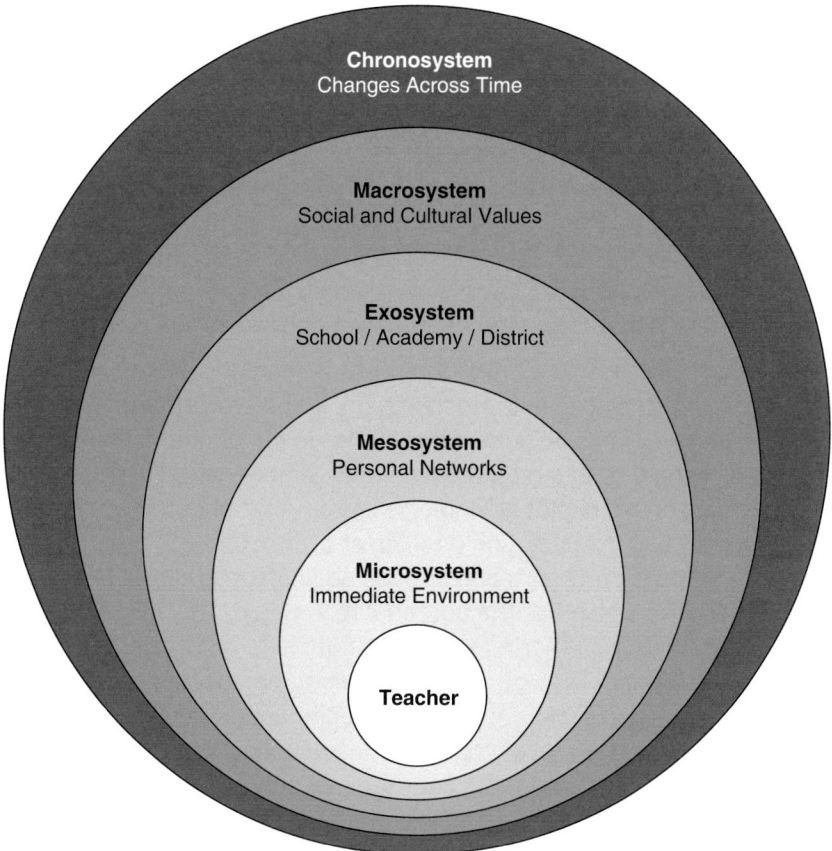

■ **Figure 12.1** Bronfenbrenner's ecological systems theory applied to teaching with digital technologies in schools

places that are created and what that allows us to experience. As Wellington comments, 'ICT can provide learning experiences that other learning strategies cannot' (Wellington, 2005, p. 32). If we accept the premise that when we're looking into our screens we are experiencing new places, then the way these places shape our experience shapes our learning.

The role that digital technologies can play in teaching science is through 'augmenting and connecting proven learning activities' (Luckin et al., 2012, p. 63). They cannot provide a 'silver bullet', neither can they replace the teacher in the classroom, but they have the capacity to 'catalyse' existing practice. This notion of technologies as a catalyst can be a powerful way in which to understand their role within the classroom. If we think of learning as a reaction among pupils, teachers, and content then the technologies are there to provide novel pathways through which to achieve the same ends, pathways which can require less 'activation energy' than others (Huppert et al., 2002). But before we turn to the classroom, let us explore what theory has to say about the role of digital technology in education.

THEORETICAL APPROACHES TO DIGITAL TECHNOLOGY INTEGRATION

This section gives a brief overview of some theoretical perspectives on digital technologies, teaching, and learning.

RAT

The RAT model (Hughes et al., 2006) breaks technology use down into three categories of use: Replacement, Amplification, and Transformation.

At the **Replacement** level, technology stands in for its non-digital equivalent. For example, instead of using textbooks pupils simply read a digital copy. This tendency to simply use digital technologies as a replacement for physical items is both prevalent and longstanding (Fraillon et al., 2020).

At the **Amplification** level technology is adding something different to the learning experience which has the potential to enrich it. This might be from the teacher perspective, e.g. sharing digital resources is quicker and easier than physical books. It might also be from the learner perspective, e.g. videos and animation in digital textbooks can enhance pupils' understanding.

At the **Transformation** level technology is enabling new forms of learning that would be impossible without it. For example, groups of pupils can comment on and annotate digital textbooks across the classroom at the same time, creating levels of interactivity which would otherwise be impossible because the digital place is not physically constrained.

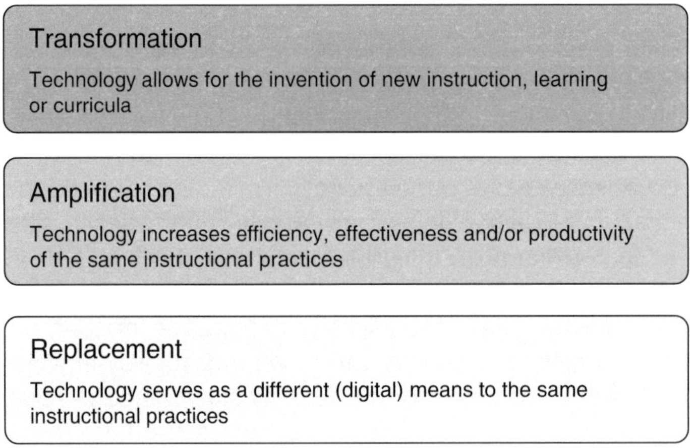

Transformation
Technology allows for the invention of new instruction, learning or curricula

Amplification
Technology increases efficiency, effectiveness and/or productivity of the same instructional practices

Replacement
Technology serves as a different (digital) means to the same instructional practices

▪ **Figure 12.2** The RAT model

The RAT model can be a useful framework to reflect on a prospective digital technology initiative and how it will support your teaching. If you are working solely on the level of replacement, for example, it may not be worth it.

SAMR

The SAMR model (Puentedura, 2006) adds an additional level to the RAT model. The four-letter acronym stands for **Substitution**, **Augmentation**, **Modification**, and **Redefinition**. Within the model, the first two levels (Substitution and Augmentation) are grouped together as 'Enhancement', whilst the next two (Modification and Redefinition) are seen as 'Transformation'.

As with the RAT model, SAMR can be a useful way to think about the role of digital technologies in your classroom. What you are really aiming for with digital technology is at least augmentation, preferably modification or redefinition. Does the digital technology allow you to speed up the process, for example or bring pupils, teachers, and/or content together in ways which would otherwise be impossible?

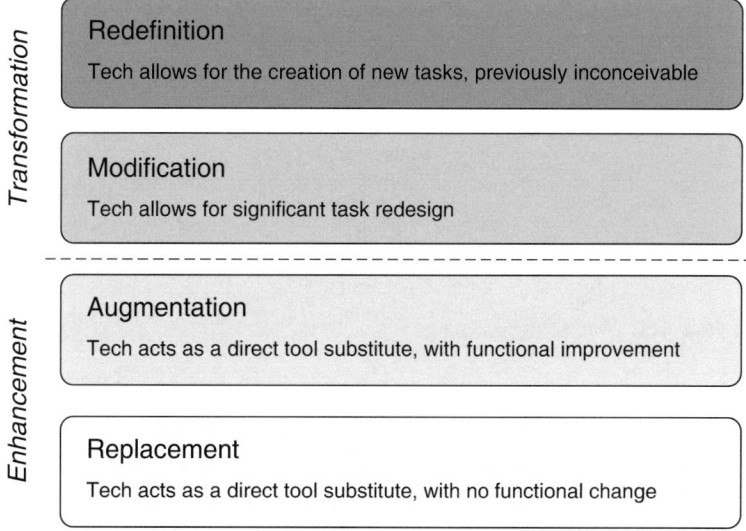

■ **Figure 12.3** The SAMR model

TPACK

The Technological Pedagogical Content Knowledge (TPACK) framework (Mishra & Koehler, 2006), is built on Shulman's (1986) Pedagogy and Content Knowledge (PCK) framework. Shulman was trying to understand what makes an effective teacher, and his original work explored the twin dimensions of pedagogy and content knowledge, arguing that you needed to have a good understanding of both to be an effective teacher. Mishra and Koehler added the third dimension of technology, arguing that to integrate technology effectively into teaching and learning you also need to have an understanding of how technology works and specifically how it can be used in relation to a specific pedagogical context.

TPACK has been shown to help student teachers learn how to make effective use of educational technologies (Canbazoglu Bilici et al., 2016).

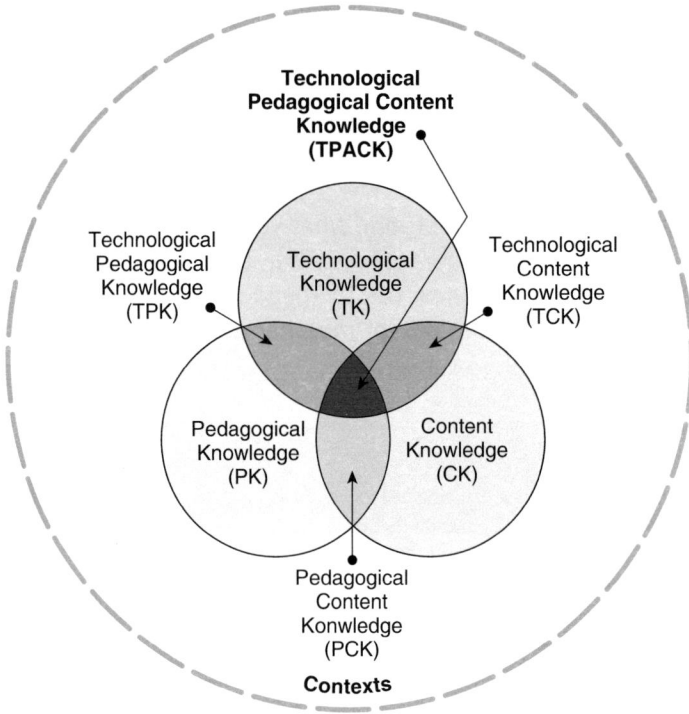

Affordance

Affordance is a cornerstone of the ecological approach. Originally coined by the psychologist James Gibson, an affordance is what something 'provides or furnishes' (Gibson, 1979, p. 127). A pencil, for example, might provide the ability to write a message on paper due to its soft carbon core but also the ability to draw attention to a specific point on a diagram due to its length and narrow end. An affordance is not fixed, it is a realistic concept which is linked to a specific context and based on the immediate needs of an individual. An affordance might be obvious to one individual yet totally opaque to another; this is especially true with digital technologies. An interactive whiteboard might be seen by one teacher asssessing a shared space for reflection, whereas another might see it as a space for distributing content. Likewise, one teacher might see a mobile phone as a distraction from learning, whereas another might see it as a valuable way in which to gather formative feedback.

The affordance concept has been widely used inssessuron (e.g. Kennewell, 2001; Conole & Dyke, 2004; Hammond, 2010) though it is not without its critics (Oliver, 2005). The term has been adopted in multiple fields but this adoption has meant much of the uniqueness of the original term has arguably been lost (Osborne, 2015). It has often been defined as the 'action possibilities' inherent in an object, but in this chapter we adopt the potentially more powerful definition of 'transaction possibilities' (Osborne, 2019). After all, when exploring a new digital technology and wondering what it will 'provide or furnish' you with, the key question is not 'what can I do with this technology?' but, more critically, 'what will I get back from using this technology?'

Task 12.1 **Evaluate your use of digital technologies**

In this task you will consider how you are currently using digital technologies in your science teaching in relation to the three theoretical approaches that have been discussed. If you are yet to start your teaching practice, consider your observations of other teachers.

1 With the RAT or SAMR approach in mind, summarise how you have used a digital technology in a recent teaching episode: at what level were you integrating technology into teaching? It may cross multiple levels.
2 With the same scenario in mind, reflect in a short paragraph how you were blending knowledge of pedagogy, content, and technology (the TPACK model).
3 Finally, with the affordance approach in mind, write down the specific affordances (or 'transaction possibilities') that you were engaging with, bearing in mind your needs and your context.

OVERCOMING BARRIERS TO EFFECTIVE USE

The ecological model which frames this chapter states that the use of digital technologies does not happen in a vacuum, but that it is shaped by factors from both the *exosystem* and the *macrosystem*. These might range from the local infrastructure in your classroom, through to policies and practices in your school, all the way up to national legislation. Understanding the barriers that you will encounter is critical for success; you cannot use smartphones for formative feedback, for example, if such devices are banned under school policy. These barriers can be thought us a hierarchy, with each resting on the other:

Digital infrastructure

Digital infrastructure refers to the technological hardware that is in your school. This might range from obvious items such as laptops and computers, to the hidden wires, cables, servers, etc. that are part of the wider infrastructure. Without a functional, resilient, and fast digital infrastructure, no digital technology use will be successful.

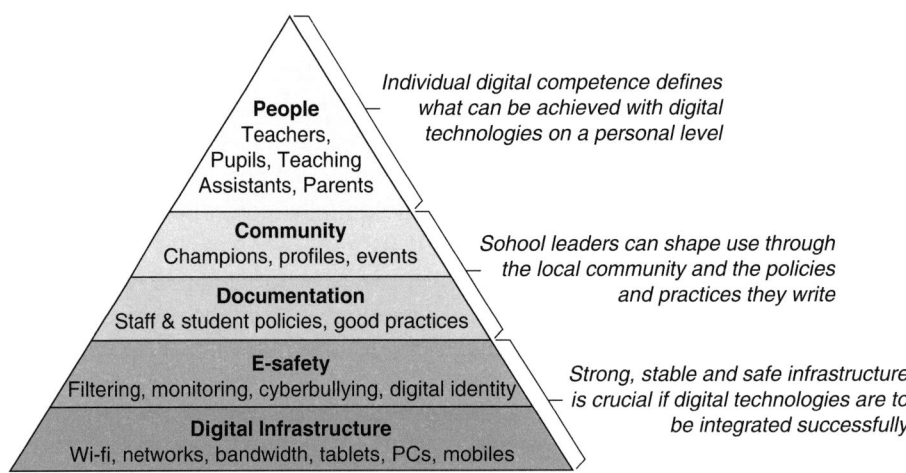

The place beyond the screen/hardware is important, not the digital technology itself. You should aim to be digitally agnostic in your choices to maximise the probability that you will be successful. There is no point becoming an expert in one specific app that is only available on iOS, for example, only to find yourself in a school full of Microsoft products (or vice versa). Focusing on specific affordances that a digital technology can provide is the key. Remember the TPACK model; a blend of Technology, Pedagogy, and Content Knowledge is required for digital technology integration to be successful.

Perhaps the biggest challenge from a hardware perspective, both in and out of the classroom, is what the pupil has. Your school may have PCs and/or tablet trolleys for use, but even if you are lucky enough to have one device per pupil, it's likely that not every pupil will have equitable access to digital technology out of school. Access may well be limited for some pupils, so you need to be careful how much you can rely on pupils working with digital technologies outside the classroom. Within the classroom you can control the way in which you use digital technologies, and this can be a boon rather than a bane. Making pupils share a device and work together, for example, can help to create extra dialogue around the tasks they are completing, enhancing the learning experience.

Finally, always have a back-up plan! Digital technologies have a habit of failing just when you need them. Fail to plan, as the old adage says, and you're planning to fail.

Policies and practices

Another challenge to using digital technologies effectively comes from the macrosystem, in terms of the social and cultural values of the school that you are in. These will almost certainly be made explicit in school policies, but there will also be local practices which may not be so clearly defined, and there will almost certainly be a mismatch between what the policies say and what happens in the classroom. You also need to be aware of the e-safety implications digital tools might present; for example, are pupils likely to be exposed to communication from third parties outside of the school and, if so, what protections are in place to shield their identities? Your IT department should provide a level of protection, but as a teacher you are ultimately responsible for the safety of the pupils in your care.

People

The final barrier to use comes from the level of digital competence of your fellow teachers and support staff but also your pupils. The term digital literacy has various definitions, but the basic concept is a measure of someone's ability to use digital technologies effectively. Don't be sure your pupils, in particular, will be experts: you may have heard of the term 'digital native', coined by Prensky (2001), which suggested that younger generations who have grown up with digital technologies simply 'get them', but more recent work (e.g. Bennett et al., 2008) has shown that they can be very poor users of digital technologies, often with specific niche expertise which does not translate well to the classroom.

Task 12.2 **Evaluate barriers**

In this task, you will discover what barriers your school might present you with when trying to use digital technologies effectively.

1. Investigate what digital infrastructure opportunities your school offers, in your department and others, both within the school and also for home use. Ensure you include levels of Internet access.
2. Find out what your school policies are regarding the use of digital equipment, for those provided by your school and also personal devices. Ensure you include e-safety policies as these may be covered separately.
3. Question fellow staff and pupils about their own attitude to using digital technologies in science; try to discover good practices and record them.
4. Sum-up your investigations in a short document, including implications for your practical teaching and ways forward.

PEDAGOGICAL APPLICATIONS OF DIGITAL TECHNOLOGIES WITHIN SCIENCE EDUCATION

Now we move into the centre of our ecological framework, exploring practical ways in which digital technologies can be used in the classroom. Evidence is provided, where possible, from empirical research, and specific apps are suggested.

Assessment

One of the keys to successful teaching is knowing what your pupils know, so you can adapt your teaching to meet the learning needs of specific classes (see Chapter 16). Digital technologies can play a role in surfacing misconceptions and there is good evidence to suggest that using them in this way has the potential to improve learning outcomes (e.g. Castillo-Manzano et al., 2016).

There are many apps which will allow you to collect formative assessment data and summarise it quickly and easily. This can help identify pupils who have not understood the topic well, as well as help pick out stronger pupils and leverage peer learning to create effective working groups.

Plickers is unique in that it does not require pupils to have digital devices, instead pupils hold up a card with a special symbol on it which can be displayed in one of four ways. The teacher uses a camera to scan the room, automatically picking up the codes and the responses of each pupil and rapidly gathering a snapshot of pupil knowledge. Furthermore, you can personalise the cards so you can track pupil's responses over time, allowing you to identify individual strengths and weaknesses.

Quizlet is a mini-quiz tool, where you can write questions yourself or pick from a bank of thousands, and share them with your pupils to test their knowledge. Its unique aspect is that it has various question formats, including standard questions, flashcards, spelling tests, and a game version. This demonstrates the technology's ability to shift the place of learning automatically, moving from one form of assessment to another based on context. **Formative**, **Kahoot!**, **Quizizz**, and **Socrative** are similar apps, each with their own twists on the concept.

Task 12.3 **Explore the pedagogical affordances of an assessment app**

In this task, you will choose a specific assessment app and trial it over a number of weeks in order to evaluate its effectiveness.

1 Pick a topic you are teaching over a few weeks, e.g. photosynthesis
2 Choose a formative assessment app and familiarise yourself with how it works. Ensure it will work in your context (note potential barriers as discussed above)
3 Find or write short quizzes; aim for one for each week or fortnight of your topic
4 Run the quizzes in classes or for homework, and use the results to inform your teaching (e.g. pair weaker and stronger pupils to support peer teaching, discover areas of misconception and reteach, stretch stronger pupils with more challenging content)
5 Evaluate the effectiveness of your digital technology integration; ask your pupils for their feedback as well.

Collaboration

The value of learning with others in education is well documented; working with peers, for example, can reveal understandings that would otherwise remain hidden. However, how digital technologies can support collaboration is trickier. Research by D'Angelo et al. (2014), for example, found that digitally enhanced collaboration between pupils did not show any statistically significant difference in impact to that without the technology.

What digital technology can provide is connections to others that otherwise would not be possible, and this can be beneficial. This might be as simple as introducing new voices through distance connections, such as a subject expert or pupils from another culture. In a well-structured learning scenario simply being able to hear and reflect on different perspectives may be enough to help pupils develop their understanding. Alternatively, because the places provided by digital technologies don't have to obey the rules of the real world, more individuals can collaborate than would otherwise be possible. As before, this will not necessarily cause more learning to take place, but it can catalyse existing practice in a well-structured scenario.

Using technologies such as **Padlet** or **Mural** pupils can collaborate on tasks in greater numbers than would otherwise be possible due to the way the digital space is experienced and hence be exposed to different ideas. These apps are effectively virtual whiteboards, but they provide much more flexibility in how you create the learning experience, including templates to structure thinking, the ability to comment on other's posts, add emotional reactions, etc. The apps are well suited to longer-term collaborative tasks, such as building shared portfolios around complex and longer topics.

Apps such as **Youtube** can be used to bring in voices from afar, pre-recorded or live, and hence extend the opportunities for discussion. These discussions can be continued and enriched using the built-in commenting systems, taking a one-way interaction into a dialogue about the topic in question. This suits 'live' topics which are often being discussed, e.g. space exploration.

There are other collaborative apps such as **Popplet** which provide the ability to co-create a type of mind map, a visual way of representing thinking which may be particularly suitable for certain topics or groups of pupils. This type of tool can be ideal for allowing pupils to demonstrate their understanding of complex topics, for example, how different aspects of a food web fit together.

Investigation

Investigating scientific phenomena can be challenging as many take place at a size or a speed which can be difficult to explore in the laboratory classroom. The growth of plants, for example, is too slow to be physically observed; molecular structures are far too tiny and outer space is far too big!

There is good evidence that the use of digital simulations can improve learning outcomes. D'Angelo et al. (2014), for example, showed a moderately statistically significant difference in impact when using simulations in science education. Osborne and Dillon (2010) note that it can help to develop conceptual improvement by reducing the 'noise' that is present in many real-world scenarios, i.e. the errors and random data that might plague a classroom activity, allowing pupils to focus on the key factors that underlie an investigation. Note, however, that there can be an inherent tension between the accuracy of a particular simulation and the noise around it. Similarly, if a simulation is made too simple then it can generate misconceptions as easily as it can help solve them.

There are many apps available which can provide useful simulations, not to mention thousands of videos of scientific phenomenon across sites such as **Youtube** and **Vimeo**. These videos can be sped up and/or slowed down to aid clarity and enhanced with formative or summative assessment tasks by combining them with apps such as **Edpuzzle** and **Vizia**.

Investigation can also be supported through digital data logging apps, which automatically record and store data in different forms, such as light intensity, acceleration, mass, etc. Many science departments will have specialist sets of data loggers which can be used to do this. Alternatively, you may be able to use smartphones. Modern smartphones contain a variety of different sensors within them, including light sensors, accelerometers, magnetometers, and microphones to name but a few, and there are many apps which can make these available for use in a science context.

Motivation

Although science is often a popular choice with pupils there will always be some who are less enthusiastic, and digital technologies can play a valuable role in increasing their motivation. Digital games are a great way to create this as they are structured experiences which can generate initial enthusiasm and direct and control that enthusiasm through their rules. This can also be achieved without digital technologies, but the use of digital tools can free you up to be more focused on pupils' actions and allows the pupils to engage in richer and more exciting forms of activity.

Kahoot! is currently one of the most popular education apps available; with its catchy theme tune and easy interface it has become very popular. Its key dynamic is to create competition between pupils, either individually or in groups. They can answer questions that you create yourself or you can pick a ready-made quiz, using a digital device within a fixed time frame scoring more points the quicker they answer. It can be useful for a flagging afternoon classroom or can be used to finish up a lesson as a form of plenary. It can be very motivating but can easily become fun for its own sake rather than contribute positively to learning, so it needs to be used wisely.

There are also apps which allow you to create ongoing rewards structures for your pupils, with badges and the like, such as **ClassDojo** and **Classcraft**. Instead of one-off use you share a collaborative online environment where pupils can track their progress against time and against each other. Parents can also be invited to join, further helping to encourage pupils; there is good evidence to suggest such parental involvement can have a positive impact on pupil achievement (Jeynes, 2005).

FUTURE THOUGHTS

We end with some thoughts on augmented reality (AR), virtual reality (VR), and artificial intelligence (AI), which are briefly defined here:

■ AR apps use a camera to overlay extra digital information onto the real world, for example, making the invisible transfer of gases and liquids during photosynthesis visible

■ VR apps use a headset to create an immersive 3D world, which can be used to create impossible experiences, for example, investigating an active volcano or the surface of Mars.

■ AI can be used to analyse pupils' knowledge and tailor content for them based on their current understanding. There is evidence that in specific subjects, e.g. Mathematics, this can improve learning outcomes.

Although AR and VR technologies are exciting and show promise, they are complicated to use effectively, and costs can be very high. Nevertheless, the type of unique experiences they provide can be valuable in science education and are well worth exploring if you have the opportunity. The use of AI in education, especially generative AI tools such as ChatGPT, is especially interesting as they can act as 'cognitive assistants', but a detailed discussion of their use is beyond the scope of this chapter.

SUMMARY AND KEY POINTS

■ Digital technologies create new digital places to explore, places which have the potential to create novel opportunities for science teaching and learning

■ Various theoretical approaches exist, including models which look at the level at which digital technologies are used (RAT/SAMR), the integration of technology knowledge with knowledge about pedagogy and content (TPACK), and the 'transaction possibilities' that digital technologies might provide (affordance)

■ There are multiple barriers to the successful use of digital technologies in education; familiarising yourself with these barriers and taking steps to overcome them is a critical step to success

■ When choosing which digital technologies to integrate into science teaching you should take a pedagogy-first approach. Focus on what it is you are trying to achieve in the classroom first, and only then seek out a digital technology that will support that focus.

Check which requirements for your initial teacher education you have addressed through this chapter.

FURTHER RESOURCES

North American Space Agency (NASA)

NASA's website includes many interactive resources for exploring the topic of space. https://www.nasa.gov/

PhET Interactive Simulations project

Free science and math simulations for teaching STEM topics from the University of Colorado, Boulder https://phet.colorado.edu/

Tech Trumps

The Tech TrumpsÆ are a curated set of digital playing cards in the top trump style listing popular digital technologies. All the techs recommended are cross-platform and are either free or include a significant free component. https://techtrumps.co.uk/

PRACTICAL WORK
Stuart Ruffle

INTRODUCTION

Science is all about trying to explain how the Universe works. To do that scientists ask questions, propose hypotheses, and design experiments. The purpose of the experiment is to test those hypotheses and ultimately answer the initial, big question. Experimental work or 'practical', as it is more widely known in schools, is an integral part of that process. The purposes of practical work are myriad, some of the most important of which will be explored in this chapter. Historically practical work in schools was largely demonstrative and instructor-led but in more modern times there has been a significant shift to it being pupil-led and investigative. Most recently the pedagogic benefits of the experimental activity itself have been highlighted as a key skill in terms of what has variously been known as 'Scientific Enquiry; How Science Works' and in the latest iteration of the National Curriculum 'Working Scientifically' (Department for Education, 2015). These changes, their impact on learning, and the promotion of STEM as a career that has national importance for the economy has attracted considerable research attention and government interest. This has been summarised in the Good Practical Science report, a piece of work by the Gatsby Foundation and led by Sir John Holman (2017). It provides an excellent overview of the state of practical work in science in education and recommends ten benchmarks, several of which relate directly to initial teacher training, student teachers, and classroom practitioners. This chapter will explore many of these ideas (with the exception of safety which is highlighted in Chapter 3) and their direct application to the student teacher and classroom practitioner.

DOI: 10.4324/9781003110187-14

OBJECTIVES

At the end of this chapter you should be able to:

■ Understand the purpose of practical work in science
■ Plan effectively for pupil progress through the context of practical work in science
■ Support pupils' understanding of the science through practical exploration
■ Develop pupils' understanding of the scientific process through effective practical work.

WHAT IS PRACTICAL?

Ask any classroom practitioner this question and you can expect to be there for a while, because when you analyse it there are many and various forms of practical work. How does a card sorting activity (probably not in most people's list of 'practical activity') differ from organising ten different sorts of nails, screws, and bolts into taxonomic groups (definitely a classic for biology teachers)? The answer is that fundamentally there is no difference. This takes us to the heart of the first issue, defining what practical work is.

Abrahams and Reiss (2012, p. 1035) define practical work as

> an overarching term that refers to any type of science teaching and learning activity in which students, either working individually or in small groups, are involved in manipulating and/or observing real objects and materials.
> (Abrahams and Reiss (2012, p. 1035)

For the teacher, it is a vehicle for promoting the progress of pupils against a set of criteria using, amongst other things, experimental work, demonstrations of principle, and genuine investigation. Those criteria are determined by history, government, and the science community (Toplis, 2015; Royal Society of Chemistry, 2022 and Science Community Representing Education, 2008) itself and they provide a framework for practical within the learning process.

Take a moment to reflect on your memory of the purpose of practical work in your own secondary science education. What do you think practical work is? What do you think it is for? When you ask pupils similar questions, especially early in their secondary education, you get a very different set of answers, all of which generally centre around the idea of it being fun and 'doing things' but beyond that their answers are highly dependent on the background that they have received in their earlier education.

This is the biggest issue that challenges secondary science teachers at the start of their career. Your starting material is very heterogeneous! Not only do you have the variability of prior pupil experience that all subjects face around that nature of the population of pupils, but there is also an opportunity and experience gap that is arguably bigger than every other core subject. The National Curriculum in England in its various forms has ensured a description of expectations around content and skills that pupils entering secondary education should have in English and mathematics. But since the complete removal of Standardised Attainment Tests (SATs) in science in 2013, their experience of science education will be very different. In Task 13.1 you will consider different backgrounds that learners bring to their secondary education.

Task 13.1 **Thinking about your starting materials – teacher and pupil perspectives**

Consider the pupil who attends a village school of fewer than 100 pupils and 5 teachers, none of whom have a science degree as opposed to one from a large, urban school with more than 500 pupils which has a science co-ordinator who has a specialist qualification and in some schools a laboratory area that would not be out of place in a secondary setting.

Using Table 13.1, organise your thoughts about the issue of heterogeneity in the year 7 (UK pupils aged 11–12) pupil population with respect to their experience of what we might classically think of as practical work. Consider this from the perspective of limitations and opportunities from the pupils' perspective and from the teacher's perspective. These reflections will be valuable when considering planning that is based around pupils' needs.

■ **Table 13.1** A structure for reflection on the impact of prior experience of practical work in school

Perspective of pupils with very limited prior experience of practical work.	Perspective of pupils with considerable prior experience of practical work.
Limitations: Opportunities:	Limitations: Opportunities:
Perspective of teachers of pupils with very limited experience of practical work.	Perspective of teachers of pupils with considerable primary school experience of practical work.
Potential barriers to learning: Opportunities:	Potential barriers to learning: Opportunities:

THE PURPOSE OF PRACTICAL WORK

Much has been researched and written about the purpose of practical work and the value of the investment of time and resource. The Gatsby Foundation report 'Good Practical Science' Holman (2017) used the outcomes of an international survey to identify five key purposes that are common to curricula intentions and implementation across the world:

A- *To teach the principles of scientific inquiry (such as controlling variables when investigating the reactivity of different metals in the Key Stage (KS)3 curriculum in England for pupils aged 11–14 where significant errors are induced if particle size is not controlled)*

B- To improve the understanding of theory through practical experience

C- To teach specific practical skills, such as measurement and observation that may be useful in future study or employment

D- To motivate and engage pupils

E- To develop higher level skills and attributes such as communication, team-work and perseverance.

Reflective analysis by the teacher on practical work undertaken in schools should always be justifiable in terms of the majority of the previous criteria because practical work is always a learning opportunity. If it cannot be justified (or arguably if just criterion D is satisfied) then is it a valuable and effective use of curriculum time and resource? Pupil surveys like that of Murray and Reiss (2005) and Sharpe and Abrahams (2020) show that pupils value the learning and understanding derived from the different elements of practical work and do not see it solely as 'fun' or a 'time-filling exercise', especially as they progress toward external assessment milestones.

From the teacher's perspective the evaluation of national projects that were introduced to promote practical work highlights that the 'doing' of practical work is actually straightforward to implement. However, it is the reflection by both teacher and pupil on the purpose of the practical work where the value is not maximised (Abrahams & Reiss, 2012).

Classroom practitioners should always ask themselves

'Why are we doing this practical?'

'Is it effective in promoting progress?'

'What are the assessment outcomes from this piece of work?'

Changes to the summative assessment protocols, following significant reformations of KS5 (aged 16–18) and KS4 (aged 14–16), have biased the purpose of practical work in favour of the first and third questions, where the purpose, methodology, or contextual use of models in practical experimentation is now more rigorously examined (see section on assessment later).

Task 13.2 **Good Practical Science – personal audit for the student teacher**

A key skill for any student teacher is to review the experience of others, share their interpretation of that review, and apply it to their own professional development. Sir John Holman's review 'Good Practical Science' (2017) is freely available. Read sections 1, 2, and 3 and from section 3 highlight the points relating directly to student teachers and early career teachers that you recognise are priorities for your development. Write a short summary of how you will plan your professional development during your training and share those thoughts with your peers.

THE TYPE OF PRACTICAL WORK

As with defining practical, categorising it is not straightforward. Here we will consider pupils undertaking the practical work rather than teacher demonstrations. There are two major divisions that can be considered – *illustrative* and *investigative* practicals. The success of the activity in achieving its outcomes depends for a large part on the selection of the appropriate type of practical (Abrahams, 2017). That judgement is a key skill for teachers, and it is informed by the needs and skills of the pupils as well as the learning outcomes that need to be achieved. Teachers should always ask themselves, what is the number one outcome that I want my pupils to achieve and what can I simplify for them so that the outcome is more likely to stand out or be achieved?

The learning that can be achieved by interacting with the physical world is not to be underestimated (Hetherington & Wegerif, 2018). Experiential learning (Kolb, 1984) has long been recognised as a powerful route to embedded understanding and experimental work in science can exploit this in a number of ways.

Illustrative practicals are largely driven by the activity being a 'means to an end' rather than the activity itself being the learning experience. A classic example would be demonstrating a principle or embedding content, for example, the test for the presence of hydrogen gas.

Pupils can be told about the test but there is something deeply satisfying and memorable (even many years into a teaching career) in hearing a salvo of crisp, 'squeaky pops' going off across the laboratory. Even that simple chemical test can be approached pedagogically from either direction. Pupils could do a confirmatory practical when they have already been told that the presence of hydrogen is indicated by the 'squeaky pop' or they could be given the method and asked to observe and describe the outcome, which is more of a discovery activity.

This reversal of the sequence of knowledge and experiment can be used very effectively in a range of situations and the sequence fundamentally alters the degree of challenge to the pupil. Some questions for you to reflect on and consider as you plan this kind of teaching are: 'Which way around in the sequence provides the most challenge?' and 'What sort of support would be appropriate for reticent pupils'?

Illustrative practicals are sometimes criticised as being 'recipe' following, however, they can be technically very challenging such as following multiple practical steps, for example, in the A-level biology classic of the root tip squash to observe mitotically active cells (Science and Plants for Schools (2022)). Where attention to detail and perseverance are key outcomes then genuine investigation may be a bridge too far.

Investigative practicals could be seen as the highest-value experiments (Akuma & Callaghan, 2018). However, posing a question to pupils and then leaving them to it is very challenging and unlikely to yield the 'expected' results or lead to a significant understanding of content or process unless a lot of foundations are already in place. Deconstructing what they have to achieve is instructive here. Essentially, they are being asked to complete a piece of research, as they are unlikely to know the answer before they start, so they are going to need to be able to do most if not all of the following:

1 Have a solid understanding of the scientific process.
2 Have the ability to convert a question to a hypothesis and then identify appropriate variables.
3 Have a sufficiently rich experience to select an appropriate investigative method.
4 Consider the safety of that method.
5 Determine an appropriate analysis of the data achieved.
6 Spot and deal with anomalies.
7 Draw conclusions.
8 Evaluate the strength of the conclusion based on the data.
9 Evaluate the validity of the experiment and method based on the data.

In the last iteration of UK GCSEs before reformation (first assessment in 2018), where coursework was still part of the assessment, KS4 pupils (aged 14-16) essentially carried out that list of tasks under medium and high control conditions with only written support provided by the examination board and carefully planned and regulated guidance from teachers. It was therefore recognised that four to five years of practical training and familiarisation was required in order to successfully attempt a full 'assessed practical' investigation presented in this way.

All of the elements listed earlier are embedded throughout science schemes of work that draw on the requirements of the National Curriculum and they are assessed in investigation-type questions in terminal examinations.

Best practice now, as it has always been, is that only one or two elements from that list are tackled and mastered at any given time. This is a core concept in teaching highlighted by Rosenshine (2012) in his 'Principles of Instruction'. The remainder of the activity needs to be simplified or given to the pupils so that there is more clarity around what the success criteria are. For example, if you want year 7 pupils (pupils aged 11-12) to investigate variation between different pieces of measuring apparatus, 5 different thermometers in a water bath is more likely to allow them to focus on accuracy and precision rather than setting up an exothermic reaction and then measuring temperature changes with different thermometers.

Ultimately when skills are embedded, genuine investigations can be undertaken and there are a large number of different practicals that are available from the historical database of assessed GCSE practicals that come fully prepared courtesy of the examination boards! Having considered some of the elements of pedagogy in practical work, Task 13.3 asks you to consider the detailed elements of a single practical.

Task 13.3 **Simplifying the task: Directing the focus.**

Consider a practical from any scheme of work for pupils aged 11–18 (for example, in England, the AQA (2018) GCSE Trilogy practical handbook) but avoid looking at the learning outcomes. Read the method only and then try to work out what the outcomes were in terms of skills (not content). Once you have does this, compare your ideas with the scheme and evaluate both. Where you find a mismatch think about the skills that pupils gain from the activity (modify your thinking) and also how the practical could be modified to make it fit your criteria.

PLANNING PRACTICAL WORK

Benjamin Franklin is credited with the maxim that is widely applicable in many areas of life, *'If you fail to plan, you are planning to fail'*! This is certainly true when considering practical work. Planning practical work is not something that has to be done from scratch on each occasion but it does need attention every time. There is a common sequence of events that most teachers will go through that fits most occasions.

Before the practical

1 You have decided that a class practical is an appropriate pedagogical approach
2 Consult the scheme of learning.
   This will either be a home grown or an off-the-shelf commercial resource. It is most likely to contain a practical that fits the bill that has the technical requirements, pupil and teacher instructions, and safety information already in place. (Safety is covered in more detail in Chapter 3 –suffice it to say that the risk assessment process is the responsibility of the teacher and departments should have a process in place that documents the fact that risk has been appropriately considered).
3 Assess the class.
   It is very important to consider your particular class and the specific needs of the pupils within it (this is a core part of the UK Teachers' Standards (DfE, 2011)). Do you have pupils with specific learning needs or medical considerations? Do you have teaching assistant support in the class, and what will they do? You will need to think about them carefully as you go through the planning process; it is most likely that you will need to adapt the standard plan.
4 Talk to your colleagues.
   Often there will be a number of different practicals that suit the purpose that are more applicable in your context. It could be that the department has resources for the alternative but not your initial choice, or that for the sake of consistency of practice (so everyone knows later that all pupils did practical X) everyone does the same version for the same purpose.

5 Practice the practical.

Early in your training and career there is a high likelihood that you will not have done that practical at all, or at least not since when you were at school – so practice. This is **vital** and experienced colleagues will do this too. Not only does it allow you to construct a better quality learning episode for your pupils, but it also familiarises you with the equipment, the process, timings, likely pitfalls, and unexpected results and gives you the time to consider the needs of the pupils and the learning outcomes that are deliverable. Talk to your technician colleagues; they will be able to source the materials you need to practice or give you better, more reliable alternatives and they will save you lots of time and effort in finessing the details.

6 Order your practical.

Most departments run an ordering system that gives technicians up to a week ahead to prepare. In some special cases you may need weeks of advanced notice e.g., seedlings, microbiological or dissection materials. In some circumstances technicians may need to prepare materials in the class or will want to support you directly during the activity.

7 Plan the presentation.

Think about how you are going to present the practical to your pupils; do they need written instruction or a storyboard/pictorial sequence, a visual presentation or a teacher demonstration? Often a combination works well. Consider whether it is too long a sequence to do in one go – there is value in 'chunking' sections.

8 Produce the additional resources.

What are the pupils going to need? For example, whole experiment writing frames, method sheets, results tables, graph paper, success criteria, exemplars of good outcomes – these need preparing in advance. Often these resources will already be in place but review them to make sure that they fit your context

9 Planning the execution of the practical.

This is a step that is often overlooked. How will resources be available in the laboratory? Will it be laid out for pupils or is selecting equipment part of the challenge? How big are the work groups going to be? Who is in those work groups? How are you going to differentiate for your different pupils? What are your and your department's expectations around tidying up?

10 Where does the practical fit in the lesson?

What is your running order for the lesson? You need to rehearse this regularly. Experienced teachers will do this when trialling a new practical or modifying an existing one. An excellent (and resource intensive) practical can be ruined as a learning experience if everyone runs out of time.

During the practical

11 Before you set them off.

Probably the most important point. Your final part of the briefing should be safety and then get several pupils to describe back to you what the first few or more steps are. You will still get 'what do I do now?' questions but you want to reduce that as far as possible and promote independence and confidence (Rosenshine, 2012). This gives you more time to focus on the pupil learning and less time dealing with administration of the practical.

12 Remain involved during the practical.

The practical should run smoothly if it has been effectively planned! During the practical remain actively involved with the pupils. Many times they will come to you for help; do not be tempted to do their work for them – just as you might rephrase a question in class, break down the practical steps for them as they do it. Do intervene when pupils are stuck and not asking for help; it is important for them to try but remember the point of the practical was to achieve the progression of a skill or understanding rather than getting the expected results. Pupils and teachers alike can be distracted by the mechanistic 'doing' rather than progressing as a scientist.

13 Remain vigilant of activity, time, and progress towards the outcomes. The organisation of the laboratory space varies widely between settings so being mobile and trying to keep your attention on the greatest number of pupils as possible is the aim. Use your ears! If it goes quiet, you know something is afoot – either a spillage or a breakage or someone is freestyling and demonstrating a 'new and exciting' way to do the practical to their lab-mates.

After the practical

14 Reflection.

Not strictly a planning task but it is important for developing your practice that post-practical you reflect not only on pupils' outcomes but your own performance especially with regard to planning going forward – this is an important implementation of the plan-do-review cycle based on Kolb's work (1984). A useful tool to support reflection of practical is the Practical Activities to Assess and Improve Effectiveness (PAAI) toolkit developed by Robin Millar (2009).

Task 13.4 **Planning practical work – have a cup of tea**

Next time you are going to make yourself the staple beverage of teachers during planning – go through the previous sequence and write yourself out a plan of how you would teach it to a year 7 (pupils aged 11–12) class. (You probably do not need to seek advice from your technician, but 'ordering' your practical requirements is only the same as going shopping for tea and checking that you have got a kettle, cup, and spoon!)

ASSESSMENT OF PRACTICAL WORK

Assessment and feedback are integral parts of all learning (Black, 2017; Black & Wiliam, 2018; Hattie 2012) and practical work is no different. The formative assessment of practical work can cover all elements of the practical from planning, physical skills, observations made, analysis, discussion, and evaluation. For example, if investigating Ohm's law, pupils could be assessed on interpreting a circuit diagram to select the appropriate equipment; assembling a circuit; circuit fault finding; reading meters accurately; recording data; graphing; and mathematical processing of data as well as literacy skills of discussion and evaluation. It can take any form just as in any other area of your practice. What is key is the learning that comes from the practical process, and this can be cemented by effective feedback that focusses on the learning outcomes and how they can be embedded or improved in future practice. Discussion around assessment and feedback for learning is discussed in Chapter 16.

It might be expected that experimental work that forms such a central strand of curriculum design and examination specifications would be embedded in the summative assessment leading to formal qualifications. In many countries this remains the case; however in England summative feedback on practical work has changed significantly. There are no longer any summative, independent investigations at KS4 (aged 13-16) or KS5 (aged 16-18; Ofqual, 2015b). In Scotland direct summative assessment of practical science marked by teachers remains in the specification (SQA, 2022). In Wales and Northern Ireland, there is summative assessment of practical work, but it is externally assessed. The context for these changes were largely based on concerns around learner independence, plagiarism, and potential effects on school accountability measures. These are discussed by Opposs (2016).

In England at KS4 (pupils aged 14-16) a new approach has been adopted. Ofqual (2015a, 2019) and each exam board recommends a series of practical activities (a minimum of 8 for the separate sciences and 16 for combined science) that pupils need to complete during their course. The techniques and practicals included were decided by consultation with the subject specialist associations. Schools are required to allow pupils to do the full range of practicals and make a declaration to that effect. The documentation required by schools is less proscriptive and for most contexts the pupil's exercise book, that is a record of their activity, is sufficient if moderation by the examination board is required.

Pupils are assessed on their understanding of the purpose and execution of practical work as well as the associated 'Working Scientifically' skills in examination questions in their terminal assessment. The questions are based on the practicals that are on the recommended list (or minor modifications in order to change contexts or assess comparative skills) and can often involve extended writing as well as data analysis and drawing conclusions.

At KS5 in England (pupils aged 16-18) the situation is more complex. It was recognised that practical skills needed more formal recognition in order to prepare pupils for tertiary level education or employment and that the 'currency' of that should be an 'endorsement' that sits alongside the A-level qualification. The Department for Education (2014) in consultation with the professional subject specialist associations and the exam boards created a series of statements (Common Practical Assessment Criteria - CPAC) that cover the core skills associated with the A-level sciences. Each examination board publishes a list of recommended (minimum of 12) practical activities that cover the CPAC criteria and issue support for tracking pupil progress against them. The judgements are teacher assessments based on direct observation and experimental written reports by the pupils which are retained for moderation. Each centre is visited every two years (largely on a subject rotation basis) to ensure that pupils are undertaking the practical work and that teacher assessments and record keeping are robust.

Practical questions based around the core practicals relating the specification content and the CPAC skills are also included in terminal examinations. Task 13.5 encourages you to put yourself in the pupil's place when answering summative questions about experimental work.

Task 13.5 **Exploring practical assessment**

In this task we will draw on typical resources available to support practical assessment in England. If you are working with an alternative set of examinations, feel free to use the appropriate alternative from your own context.

Find past papers and mark schemes for your equivalent contexts. An England and Wales example would be a GCSE (age 14–16) separate science paper in a subject area that is **not** your primary area of expertise and an A-level (age 16–18) paper that is (for A-level be sure that you have selected the paper that specifically has a high percentage of practical questions if your exam board is organised that way). Find the practical questions and answer them under exam conditions. Then use the mark scheme to mark your work twice. Use the first time of marking to assess what you got right and the second time to identify what you should have included.

SUMMARY AND KEY POINTS

This chapter has only scratched the surface of the role of practical work in science education and presented you with the basics of guiding pupils on that journey. The key points are:

■ By thinking about the purpose, planning, and assessment before you start any piece of practical with pupils you will significantly improve your chances of successfully achieving the outcomes with them and driving their progress forwards

■ Just like practical work itself, the teaching of practical work is experiential in nature and just like the pupils you will learn far more by doing than by reading about it!

Check which requirements for your initial teacher education you have addressed through this chapter.

FURTHER RESOURCES

Abrahams, I., & Reiss, M. (2017). *Enhancing learning with effective practical science* (pp. 11–16). Bloomsbury Academic.

This book goes into significant detail around practical work and includes lesson plans, worksheets, and health and safety information across a range of biology, chemistry, and physics topics.

Auty, G. (Ed.). (2015a). Practical work I. In *School science review*. Association for Science Education.

Auty, G. (Ed.). (2015b). Practical work II. In *School science review*. Association for Science Education.

Two volumes of the Association for Science Education's research journal that focus on the role of practical work in science education and all aspects related to it. It contains review articles from many leaders in educational research in the UK.

Black, P. J., & Wiliam, D. (1998). *Inside the Black Box: Raising standards through classroom assessment*. King's College.

This short pamphlet, available as a download, is the archetype for Assessment for Learning. It has been reviewed and built on many times, but I recommend the original as the starting place.

Cadwallader, S. (2018). *The impact of qualification reform on the practical skills of A level science students*. Ofqual/18/6433. Retrieved September 27, 2023, from https://assets.publishing.service.gov.uk/government/uploads/system/uploads/attachment_data/file/747471/6433_FINAL_-_A_level_science_study_4.pdf

This document appears to be a UK government review of a policy change for England but it has significant detail in it for guiding your thinking about the relationship between practical work and its ultimate assessment at A-level. It uses statistical research methods of examination data to investigate where pupils struggle in examinations (and therefore where ground is to be made)!

Hattie, J. (2012). *Visible learning for teachers*. Routledge.

The teacher-specific version of the seminal work in the current age on maximizing the impact of learning to drive pupil progress.

Holman, J. (2017). *Good practical science*. The Gatsby Charitable Foundation. Retrieved September 27, 2023, from https://www.gatsby.org.uk/education/programmes/support-for-practical-science-in-schools

An extensive review of practical work in schools in the UK. It covers the purpose of practical work and develops a series of benchmarks that form a framework for quality experimental work in schools. There are a number of companion appendices that include the UK school survey, comparisons to international curricula, and a review of metastudies of practical work pedagogy.

LANGUAGE AND LEARNING SCIENCE

Judith Hillier

INTRODUCTION

This chapter explores the importance of language in learning science and some of the challenges that language may present to pupils in their learning of science. Language may be something you have taken for granted so far in your professional development as a science teacher, but it can pose some very real difficulties for teachers and pupils. Reading scientific texts, writing using different scientific genres, and classroom discussion can all be challenging experiences for pupils and hence barriers to learning. These issues are discussed in detail in this chapter to help you to develop your knowledge and understanding of the role language plays in learning science and the difficulties language poses to many pupils in their science lessons. There is also an in-depth exploration of approaches you can use in your classroom practice to help pupils to develop their literacy and oracy skills and hence support their learning in science.

OBJECTIVES

At the end of this chapter you should be able to:

- Recognise the ways in which language plays a role in learning science
- Understand some of the challenges language can pose to pupils as they learn science
- Consider a range of strategies to support pupils' talking, reading, and writing in science.

THE IMPORTANCE OF LANGUAGE IN LEARNING SCIENCE

Many people associate learning science with practical work and mathematical calculations, and these are both fundamental parts of learning science, but science and learning about science rest on the foundations of language (Sutton,

DOI: 10.4324/9781003110187-15

1992). We think of science as the study of all things biological, chemical, and physical – describing and explaining the world around us, but this needs language. If I study a frog, I can find out about its lifecycle, habitat, diet, and behaviours, but if I call it a 'leggy-thing' because I don't know the term 'frog', this will severely hamper my ability to learn about it and to share what I have learnt with others. Figure 14.1 depicts Kepler trying to explain his laws of planetary motion, but his listeners don't understand the words he is using, never mind the concepts he is trying to convey. Language is important in learning science because teachers and their pupils need to have a shared understanding of words in order for pupils to learn (Lemke, 1990).

▪ **Figure 14.1** Johannes Kepler's uphill battle, illustrating the need for shared language in order to develop conceptual understanding

Source: Copyright, 1980 by Sidney Harris, The American Scientist Magazine, reproduced with permission

Neither should pupils just learn set phrases:

> We want them to be able to construct the essential meanings in their own words . . . but they must express the same essential meanings if they are to be scientifically acceptably . . . that is what we mean when we say we want pupils to 'understand concepts'.
>
> (Lemke, 1990, p. 91)

This is the challenge for science teachers: to get pupils to use scientific language appropriately but without rote-learning it.

THE NATURE OF LANGUAGE IN SCIENCE

Having established the importance of language in learning science, what are these scientific words that are so hard for pupils to learn and to use? There are a number of frameworks used to group vocabulary, which will now be discussed.

One framework is the *tiers of vocabulary* model, developed by Beck et al. (2013; Figure 14.2), a helpful way of thinking about which words may need to be explicitly taught to pupils: if you are teaching pupils for whom English is an additional language, Tier 1 words may need introducing. For Tier 2 words, explain what this word means in the context of a science lesson (which may be different to how it is used in other subjects). Pupils will need you to tell them how to pronounce Tier 3 words, as well as explain the meaning, and they will need to practice spelling these words.

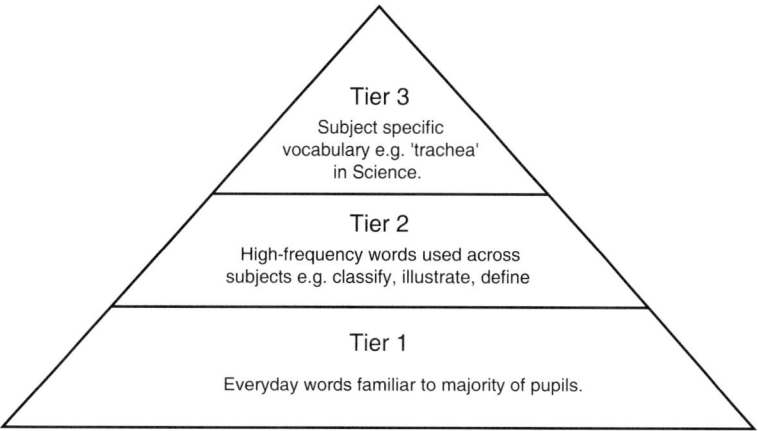

■ **Figure 14.2** Tiers of vocabulary model

Source: adapted from Quigley and Coleman (2019)

Another framework is that of scientific, technical, and non-technical words, discussed by Wellington and Osborne (2001). Table 14.1 shows that every science lesson involves some of these words. Wellington and Osborne advocate that you should be aware of the language you are using in your classroom, you should try

■ **Table 14.1** Scientific, technical, and non-technical terminology used in science lessons

Scientific terminology		Semi-technical terminology		Non-technical terminology which is widely used in science	
Unique to science	*Used in everyday language as well*	*Single meaning*	*Multiple meanings*	*Single meaning*	*Multiple meanings*
Voltmeter	Reaction	Emit	Heavy	Maximum	Independent
Diffraction	Current	Repel	Map	Relevant	Rate
Electron	Force	Particle	Attract	Limit	Complex
Photosynthesis	Family	Flow	Relative	Factor	Initial

Source: adapted from Wellington & Osborne, 2001, p. 18)

to be clear and consistent in your use of language, and you should take time to explain the meaning of the language you are using.

The final framework to consider is the taxonomy of words in science (Box 14.1), again taken from Wellington and Osborne (2001). This shows how some words in science are closely associated with objects or processes, some of which are concrete realities that can be brought into the classroom and others much more abstract, particularly the more theoretical concept words. Pupils often first learn the words associated with concepts such as 'atom' before understanding the concept of an atom. As a science teacher, the word 'atom' will probably conjure up for you a mental image of an atom with its nucleus and electron orbitals; you may remember the plum pudding model, Rutherford's experiment, and the Bohr model of the atom (See Chapter 6 and Chapter 10). All of this is part of your concept of an atom, developed over many years; your pupils are beginning this learning, and you are there to help them. This use of a single word or phrase to represent a complex idea is called 'nominalisation', for example 'the bending of light as it crosses a boundary between material with different densities' becomes 'refraction' (Halliday & Martin, 1993, p. 66).

Box 14.1 **A taxonomy of words in science (adapted from Wellington & Osborne, 2001, p. 20)**

Level 1: Naming words

■ Familiar objects, new names (synonyms) e.g. trachea for windpipe
■ New objects, new names e.g. Bunsen burner
■ Names of chemical elements e.g. potassium
■ Other nomenclature e.g. vertebrates and invertebrates

Level 2: Process words

■ Capable of being shown e.g. combustion, distillation
■ Not capable of being shown e.g. evolution

Level 3: Concept words

■ Derived from experience (sensory concepts) e.g. colours, density (think floating)
■ With everyday and scientific meanings e.g. current, power, salt, fruit
■ Theoretical constructs e.g. element, mass

Level 4 Mathematical 'words' and symbols

Why consider all three frameworks – isn't one sufficient? Language is complex, with no single way to examine it. Each framework acts like a microscope lens with different levels of magnification; each reveals something of the nature of the language used in science lessons; changing from one to the other changes the level of detail, still providing valid observations with implications for you as a classroom teacher. Having discussed the importance of language in learning science and its nature, it is now time for you to consider your own use of language in lessons in Task 14.1.

Task 14.1 **Evaluating your own use of language**

During your school experience, reflect on your use of language in a lesson, (or observe another teacher's use of language) and consider the following questions:

1 What new words did you introduce in the lesson?
2 How did you introduce and explain new words? Were these scientific/semi-technical/non-technical words?

Discuss your reflections with your mentor, tutor or another student teacher – what are the implications for your classroom practice as a teacher? How might you take account of these issues in your lesson planning? Take reflective notes which you may wish to add to your portfolio of evidence if you are required to keep one.

Having considered the nature of the language used in science lessons, it is now time to explore some of the challenges pupils experience with language when learning science.

CHALLENGES EXPERIENCED BY PUPILS LEARNING SCIENCE

Learning the language of science can be compared to learning a foreign language. At some point during your own schooling, you probably learnt a new language such as French, German, or Spanish (the ones most commonly taught in UK schools). You may remember struggling to pronounce words and to remember their spelling, never mind understanding what you were reading or hearing! But over time, with lots of practice, words like 'maison' or 'Brot' became familiar words to read, write, and say, and you knew that it is 'la maison' and 'das Brot'. Eventually, you may have travelled abroad and successfully used this new language in a restaurant or a hotel, but it all started with a teacher helping you to associate 'maison' with 'house', and 'Brot' with 'bread', which were concepts you already knew from everyday life. This is the first challenge for pupils learning science: they might 'learn' a new scientific word, but what concept, what mental model, is it associated with? (See Chapter 10: Constructivism for an introduction to the idea of mental models and schemas). In Box 14.2, you can read two short excerpts from interviews a student teacher carried out for an assignment. Pupil 1 associates the process of evaporation with the idea of water escaping; as a mental model, this could be an acceptable starting point for thinking about evaporation, but it runs the risk of personification whereby the pupil starts to think of the water *wanting* to escape. It is also easy for a pupil who is verbally using 'escaping' as an equivalent term for 'evaporation' to write 'escape' instead of 'evaporation', an error likely to be penalised in written assessments.

In contrast, pupil 3 uses the word 'evaporation' but has a very muddled mental model of what it means - are the particles dying or leaving or something else? Pupil 4 associates evaporation with the water cycle - this is problematic for two reasons: they may think that water is the only substance that can evaporate, and they appear to think that the whole of the water cycle is evaporation, when it also involves condensation, precipitation, and collection. These mental models are partial and prone to error. As science teachers, being aware of the flawed mental models pupils are likely to associate with key scientific words (sometimes called misconceptions) is an important part of our professional knowledge, as then we can address these in lessons. A good starting point for learning about common misconceptions is *Making Sense of Secondary Science* (Driver et al., 2014). See Chapter 10.

Box 14.2 **Pupil interview responses to a scenario about evaporation, with thanks to Harry Gilchrist.**

A PGCE student interviewed pairs of pupils for an assignment, using this scenario: "*You pour a drink of water and leave it in the sun on a windowsill in a locked science lab. After three hours you return and discover that the water level in the glass has gone down.*"

PGCE student: "So what has happened?"
Pupil 1: "The water has got hot and turned into steam."
Pupil 2: "It's evaporated."
Pupil 1: "Yes it's escaped?"

PGCE student: "So what has happened?"
Pupil 3: "Evaporation."
PGCE student: "Nice scientific word, what is that?"
Pupil 3: "it's where the particles die, or leave I don't know."
Pupil 4: "Evaporation is where water heats up and turns into steam goes up into the clouds then rains when

---

As well as the challenge of learning what scientific words actually mean, pupils also experience challenges reading, talking, and writing in science lessons. Reading in science lessons is challenging because of the nature of the texts we commonly use. Three frameworks for analysing the difficulty of vocabulary used in science lessons have already been discussed (Figure 14.2, Table, 14.1 and Box 14.1), but the sentences using this vocabulary are also often quite complex. Science is about building up descriptions and explanations of the world around us, so scientific language connects scientific concepts together in careful and precise ways. The use of connectives in sentences increases the complexity of the text, as does the need for precision. For example, electrical current could be described as *'moving charge'*, but this wouldn't be acceptable to most physicists, so we teach that *'current is the rate of flow of charge'*. Comparing these two phrases, it is easy to see why pupils would find one much easier to read than the other. Historically, science textbooks have had a high reading age, but much has been done to improve the readability of texts, whether in books on worksheets or on websites: using ideas from cognitive load theory, appropriate layouts, careful use of fonts, and judicious use of clear and relevant illustrations all help (e.g. Nyachwaya et al., 2016). Useful advice about dual coding – using both words and pictures – is widely available (e.g. Caviglioli, 2019), although care needs to be taken to integrate these so as not to overwhelm learners, particularly lower attaining pupils (Corradi et al., 2014). Nevertheless, explanations still need to be built up, and long words still need to be used.

Another challenge is talking in science lessons. Mortimer and Scott (2003) write about the centrality of talk in the process of making meaning – relating existing ideas to new ideas – in order to learn, but this poses challenges for both pupils and teachers. For some pupils, the thought of saying something out loud that might be wrong, using words that are difficult to pronounce, can be intimidating, and it is much easier to say 'I don't know'. For teachers, it is relatively easy to ask short, closed questions, initiating a short response from a pupil which can then be evaluated, called an Initiation-Response-Evaluation (I-R-E) pattern of discourse (Mehan, 1979). Examples would be 'What's the chemical symbol for oxygen?' or 'What is the gas produced during photosynthesis?' Whilst these facts are useful in pupils' learning of science, these are very much lower-order questions, and we want our pupils to develop much deeper understanding. Dialogic teaching has been espoused as a collaborative and supportive way to harness the power of talk in the classroom (Alexander, 2020).

The final challenge is that of writing in science lessons. One barrier here is that many pupils think they shouldn't have to write in science lessons, but if you look at their books for English or history, you will see that they have written pages of text, so as a teacher, the challenge is to make writing a normal part of science lessons. It is important that this is actual writing, not mere passive copying down (as per the well-known saying that lectures are where the notes of the lecturer are transferred to the notes of the pupil without going through the mind of either). And the idea that pupils need their notes for revision is fallacy – revision guides and the plethora of online resources are much easier to read than their own handwriting, especially if they frequently make mistakes when copying from the board or textbook. Writing in science uses different genres to other subjects: there are experimental accounts, explanations, descriptive reports, and scientific arguments (Martin & Miller, 1988). Writing is usually in the passive voice, which pupils are less familiar with, and the scientific vocabulary and technical words have to be used and spelt correctly. All of these are skills which have to be taught and learnt, and for pupils who are nervous about getting things wrong, it can be much easier to write very little at all.

All of these challenges become much more daunting if, as a pupil, you are also in the process of learning English as an additional language (EAL), and you can explore this in Task 14.2.

Task 14.2 **Shadowing an EAL pupil**

As part of your school experience, ask your mentor to help you organise shadowing an EAL pupil for a day (or for as long as possible), and record your observations using these questions to guide you (think about reading, talking, and writing):

1 What issues did the pupil face in lessons?
2 What support did they receive?

Now consider what issues an EAL pupil might face in your lessons and discuss with your mentor what additional support you could put in place for these pupils. Keep these notes in your portfolio of evidence.

As a science teacher, you want your pupils to understand and use scientific words confidently and accurately in their talking, reading, and writing in science lessons and in the future, and the next three sections will discuss ways you can support your pupils with this.

SUPPORTING PUPILS' READING IN SCIENCE

To start with, pupils will be more encouraged to develop their reading skills in science if their environment is conducive to reading: a well-lit, relatively quiet room where all the pupils can easily see the text which is presented in an attractive way will help. Your own attitude towards reading as a teacher will also help – not false positivity but talking about science books you have read, having science books and magazines in the classroom, and never using reading as a punishment or forcing someone to read aloud – these all help to set the tone.

The next thing to consider is the design of the task: if you ask most pupils to read a text and make some notes, they will struggle to do anything constructive – you need to engage them in *active reading*, with a clear and specific purpose, appropriate support, and in a collaborative way. These strategies are often called *directed activities related to text* (DARTs) and you can find numerous examples on the internet. There are two main groups of DARTs: reconstruction activities where the teacher has modified the text and the pupils take a problem-solving approach to complete it and analysis activities using unmodified text where the pupils initially find specific information and can then use this to answer questions or construct a new table or diagram (Box 14.3). In Task 14.3, make a DART for a lesson you will be teaching soon.

Box 14.3 **Examples of directed activities related to text (DARTs)**

Diffusion

Diffusion is when one gas or liquid mixes with another gas or liquid. For example, when you pour milk into tea, initially all the milk particles are close together – they are in an **area of high concentration** of milk particles. Particles in liquids and gases move about randomly, changing direction when they collide with each other. Over time, this movement results in the milk particles moving to an **area of low concentration** – they spread out throughout the tea. You can, of course, speed this up by stirring the milk into the tea! Diffusion is how substances, such as oxygen, move in and out of cells in living things.

Reconstruction DART

1. Using the text about diffusion above to match the endings of each sentence to the correct beginning.

Beginnings	Endings
Highly concentrated coffee molecules	lower concentration than they started in.
Coffee molecules begin to	enter the cup of hot water.
Coffee molecules are now in a molecules.	spread out in between the water

2. Now draw a particle diagram to match each completed sentence.

Analysis DART

Use the text about diffusion above to answer the following questions:

1. Describe the movement of particles in a gas or a liquid.
2. Why do particles in a gas or a liquid change direction?
3. What do we call this process of particles spreading throughout a container?
4. Explain how this process helps us to get rid of the waste carbon dioxide produced in our cells by respiration.

Task 14.3 **Making your own DART**

Choose a topic you will be teaching in the next few weeks and find a text about it (this could be a textbook or website). Now make a reconstruction DART and an analysis DART for this text. Share these with your mentor or the class teacher, and discuss how you will give the pupils the text, what instructions you will give them, how might you help them collaborate, and what will you do as a follow-up activity.

   Once you have taught the lesson, evaluate the effectiveness of this activity – what went well and what would you do differently next time. Keep these notes in your portfolio of evidence if you are required to keep one.

SUPPORTING PUPILS' TALKING IN SCIENCE

Managing classroom discussion effectively can feel very daunting as a beginning teacher – nothing is more disconcerting than asking a question (particularly one you think is easy) only to be deafened by the silence when no one responds, and equally alarming is the thought of setting the whole class talking and then not being able to regain their attention. Most classroom discussion is about answering questions, but research has shown that usually it is teachers asking the questions, waiting for a very short time and then pupils giving one- or two-word answers (Evagorou & Osborne, 2010). As discussed earlier, a diet of rapid-fire closed questions with monosyllabic answers is unlikely to lead to deep understanding. Chin (2007) conducted a rich study into teacher questioning that stimulated pupils to think and talk productively. Key features included purposeful questioning where teachers found out what pupils think and encouraged them to expand on their ideas; questioning sequences that involved multiple pupils; questions that were adapted in light of pupils' responses; more open questions that required higher-order thinking and longer answers; teachers gathering several responses rather than immediately telling pupils whether their response was right or wrong, and involving all pupils in evaluating the responses. All of these approaches encourage pupils to talk and so to become more confident and accurate in their use of scientific language.

So how would you achieve this in the classroom? Well, there are multiple ways, and much will depend on the relationship you build with the pupils you are teaching. Purposeful questioning which explores pupils' thinking can be stimulated by a carefully chosen photograph where you ask the pupils to think for one minute about what is going on, to talk in pairs for one minute, and then to share their ideas and explanations using hands-down questioning. Concept cartoons can also be used (Naylor & Keogh, 2000). This much-researched tool uses a simple cartoon to present pupils with an everyday scenario about a scientific concept and a variety of possible opinions about it, including common misconceptions, and the teacher can set up a discussion with the pupils (using think, pair, share for example) about which opinion the pupils think is best. Follow-up activities are also suggested where the pupils can test out the opinions in order to reach a conclusion which is closer to the scientific viewpoint.

Questioning sequences can be set up using *pose, pause, pounce, bounce* (or Tigger questioning as I have nicknamed it): you pose a question and pause, giving pupils time to think. This is important, because if you expect an answer immediately, you will only get lower-order answers – it takes time to think of the words to use and to construct sentences to answer a question such as 'How would you classify these different plants?' or 'What outcome would you predict for this experiment if we changed this variable?' Having given all the pupils time to think of an answer, you choose one pupil and ask them to share their ideas – if you choose the pupil before asking the question, the rest of the class know they don't have to bother thinking about the answer – and then you bounce the question around the classroom: 'Do you agree with this answer?', 'Could you add more detail to this answer?', 'That's good, but could you put that answer into more scientific language?' This will help you to see what various pupils are thinking rather than just one, will give multiple pupils the chance to practise using the scientific language, and will enable the class to co-construct a scientific answer using the pupils' ideas and words – giving them ownership will help them learn.

This sort of class discussion doesn't just happen though; it requires you to plan your questions, good questions that will stimulate productive thinking as well as those snappy recall questions that remind pupils of the key words being used in the lesson. One way to plan questions is to use Bloom's Taxonomy, which dates from 1956 and is a taxonomy of educational objectives. Whilst there are debates about its use (e.g. Pring, 1971; Colder, 1983; Newton et al., 2020) and there are other models of lower-order and higher-order thinking, resources like the Bloom's Teacher Planning Kit (easily found in a search engine e.g. https://www.academia.edu/30701945/Blooms_Taxonomy_Teacher_Planning_Kit) are a really useful source of question stems. Good class discussions also require setting clear expectations for an appropriate classroom atmosphere, where pupils listen to each other with respect, where having a go is more important than getting it right the first time, and making a mistake is an opportunity to learn.

Now it is time for you to consider the types of questions you are asking in lessons and how you are managing this questioning in Tasks 14.4 and 14.5.

Task 14.4 **Types of questions being asked**

During your school experience, make a note of the questions you are asking in lessons (or as if your mentor can do this for you in a lesson observation). Discuss with your mentor or tutor how many of your questions are closed, lower-order ones (one- or two-word answers) and how many are open questions, requiring higher-order thinking. If there is not time to discuss this with someone else, reflect on these questions for yourself. Try planning some questions to use in a lesson, and afterwards evaluate what went well and how you could develop in the future.

Task 14.5 **Managing questioning**

During your school experience, spend a few lessons observing teachers and their questioning:

1 Where does the teacher stand to ask questions? Do they move around?
2 Which pupils are asked questions?
3 What sorts of questions are being asked e.g. open, closed?
4 How do pupils respond to these (e.g. hands-up, hands-down, mini-whiteboards, written answers)?
5 How does the teacher respond to pupils? Observe how they praise the pupils and how they respond to wrong answers.

6 How long does a period of questioning last and when in the lesson does it occur?

7 If you have time, note down a sequence of questions to see how teachers build up a sequence of questions.

Now look at the plan for a lesson you are going to teach soon. Have you made decisions about how you will manage questioning? What additional information might you add to your lesson plan? Discuss this with your mentor if possible, and keep your notes in a portfolio of evidence if you are required to keep one.

SUPPORTING PUPILS' WRITING IN SCIENCE

Just as pupils need to be actively engaged in reading about science, so they need to be active participants in writing in science, gaining familiarity with scientific vocabulary and with genres of writing in science. The use of key word lists and glossaries can help pupils to practice writing out the words, and recap activities such as crosswords and wordsearches can be useful reminders of spellings. The next step is encouraging pupils to write their own sentences: at the end of a class discussion, ask them to write one or two sentences summarising what has been agreed. Similarly, ask pupils to write a prediction and explanation for the outcome of an experiment – this can then be discussed and revised afterwards. Sentence starters support pupils with this, giving pupils a sense of the 'voice' they are meant to be using. For longer pieces of writing, this can be extended using writing frames (see Task 14.6), and again there are many examples on the internet. Be explicit about your expectations: give pupils a model answer, ask them to identify the key features, and agree a set of criteria which they have to meet before spending a long time on a written task; this will really improve the quality of the work pupils produce.

Task 14.6 **Writing frames**

Look at the writing frame below for an experimental account. Now make your own writing frame to use in a lesson. After the lesson, discuss with your mentor what went well and what could be improved in future. Keep these notes in your portfolio of evidence.

Aim
What were you trying to find out in the experiment? Was there a hypothesis?

Methods
What equipment did you use? What steps did you follow? What safety precautions did you take? What were the variables (dependent, independent, and control)?

Results

How did you record your results? Should you draw a graph (what sort of graph)? Think about labels, units, intervals between data points.

Conclusions

What conclusions can you draw from the data? What are the limitations of the experiment? How could you improve it?

SUMMARY AND KEY POINTS

This chapter has covered the following key points:

- Learning the language of science is a vital part of learning science
- Language in science is distinctive and complex
- Pupils need teachers' support to understand and use scientific language appropriately in reading, writing, and talking about science
- Now check which requirements for your initial teacher education you have addressed through this chapter.

FURTHER RESOURCES

Education Endowment Foundation (2019). *Improving literacy in secondary schools: Guidance report*. Retrieved September 28, 2023, from https://educationendowmentfoundation.org.uk/tools/guidance-reports/improving-literacy-in-secondary-schools

This guidance report from the Education Endowment Foundation summarises the best available evidence on how to improve literacy in secondary schools. Although not specific to science, there are lots of helpful suggestions both for classroom teachers and school leaders.

Mortimer, E. F., & Scott, P. H. (2003). *Meaning making in secondary science classrooms*. Open University Press.

This book is also fairly short and easy to read and uses a range of theoretical ideas and real-life classroom examples to explore how to make effective use of classroom talk to develop pupils' understanding of science and their science communication skills. Again, a book worth reading and then revisiting to think about what pupils in your classrooms are saying.

Secondary National Strategy. (2009). *Developing writing in science*. Retrieved September 29, 2023, from https://www.stem.org.uk/elibrary/resource/31727

These resources, now hosted by the STEM Learning library, were developed as part of the Secondary National Strategy project in the 2000s and provide detailed guidance and examples for science teachers working to develop their pupils' writing in science.

Thinking Together Project Online. Retrieved April 26, 2021, from https://thinkingtogether.educ.cam.ac.uk/

This website showcases classroom-based research and resources for teachers developed by a research team led by Professor Neil Mercer at the University of Cambridge on a dialogue-based approach to the development of children's thinking and learning. The resources are particularly helpful for guidance about how to set up group work effectively.

Wellington, J., & Osborne, J. (2001). *Language and literacy in science education*. Open University Press.

This short and highly readable book draws together research and good practice to discuss language in the science classroom, with a number of useful practical implications for teachers. Many of the ideas discussed in this chapter are explored in more detail and it is a valuable book to read and re-read throughout your teaching career.

MATHEMATICS IN THE SCIENCE CURRICULUM

Victoria Wong and
Luke Graham

INTRODUCTION

There are few who would argue that science is dependent on mathematics. Which aspects of mathematics varies between science subject, but all rely to at least some extent on numerical arguments, data, or techniques.

This importance of mathematics is increasingly being recognised in the science curriculum. The 2015 science curriculum in England prescribes that in the GCSE science examinations at age 16, 20 per cent of the marks must be for specified mathematics in the context of science (10 per cent in biology, 20 per cent in chemistry, and 30 per cent in physics; Ofqual, 2015b). Two of the eight practices of science of the US Framework for K-12 Science Education (NGSS, 2023), are mathematical: analysing and interpreting data and using mathematical tools. Osborne (2014) argues that mathematics in science has both a communicative function (graphs, for example, to communicate experimental findings) and also a structural function allowing for logical deduction (for example understanding the implications of surface to volume ratios). Mathematics in science is not simply a tool for the analysis of data but a way of thinking and reasoning.

A key aim of science education, therefore, should be that pupils are able to think and reason and communicate mathematically about scientific ideas that are quantitative. Many pupils, however, find using mathematical ideas in science very difficult and a significant proportion are scared of mathematics. Despite this, research in the UK (Goldsworthy et al., 1999) and elsewhere (Turşucu et al., 2017) suggests that science teachers often do not see teaching about mathematical ideas to be part of their role, even when those ideas are crucial to understanding science. Instead, they believe that pupils' difficulties with mathematics should be solved by the mathematics department. The aim of this chapter is to demonstrate that it is part of a science teacher's role to support pupils in using mathematics within science and to suggest ideas about how to promote mathematical thinking in the science classroom.

DOI: 10.4324/9781003110187-16

OBJECTIVES

By the end of this chapter you should be able to:

- Understand why science teachers need to teach about mathematics within science
- Appreciate some of the difficulties that pupils have with using mathematics in science
- Be confident in planning to teach lessons which include a mathematics in science component
- Consider ways to support pupils to become more competent and confident in dealing with numbers in science.

MATHEMATICS IN SCIENCE: SOME CHALLENGES

Language

There are differences in the way that mathematics is used in mathematics itself and in science. In an ideal world there would be consistency in terminology and approaches between the two subjects. However, mathematics and science are different disciplines with different aims and practices which inevitably leads to some inconsistencies. It is helpful if science teachers are aware of these as they can cause difficulties for pupils and frustrations for teachers who do not understand them. For example, it is quite a common task in both science and mathematics to ask pupils to draw a 'line of best fit' on a graph. Science teachers who have carefully taught pupils how to draw a freehand curve are often aghast when they automatically reach for a ruler in later lessons. However, in mathematics a line is, by definition, always straight, so using a ruler there would be the correct approach. The apparent similarity between the tasks but with an expectation of different outcomes can cause problems for pupils and teachers.

The Association for Science Education have a useful (and free) publication *The language of mathematics in science* (Boohan, 2016) which aims to help teachers understand the differences in language between the two subjects.

Science teacher expectations

An issue often encountered is that pupils may require mathematical ideas, tools, and techniques in science before they have been covered in their mathematics lessons. When science teachers expect pupils to have skills that they do not have then it can lead to lessons going awry as this teacher describes:

> I was talking to a science teacher recently who was doing a lesson on Hooke's Law, which is where you put masses on a spring and then you measure the

spring. And she went, 'So my learning intention was for them to learn that the stretch in the spring is proportional to the force', but she said, 'right, okay, so the first challenge was none of them could measure the spring properly, because they weren't using the zero on the ruler. Then loads of them had issues drawing the axes on their graph and doing the scale properly, then plotting the graph, then drawing the line of best fit'. So she said 'my whole learning intention just got completely lost in all this maths that they were struggling with'.

(Quoted in Wong, 2018, p. 135)

This description sounds familiar to anyone who has either observed or taught younger pupils in a busy lesson aiming to collect some data from a practical activity and then graph it. The science teacher expects that pupils can draw a graph from scratch; that they cannot hinders their science learning. The easiest response from the science teacher is to blame the mathematics department for not teaching the pupils what they need – and this response is frequently seen (Wong & Dillon, 2019). However, it could also be argued that the teacher had too many learning objectives for one lesson and their expectations were simply too high for most of the class. It is important that science teachers think carefully about whether the expectations they have are realistic or whether they need to adapt to the realities of what the pupils they are teaching can actually do. This is as true for mathematical skills as it is for any other aspect of teaching science. The task could be simplified by, for example, providing axes and scales for most pupils and expecting them to just plot the points and draw the line of best fit.

Differences between mathematics and science

There are fundamental differences in the use of mathematics across the two subjects that go beyond language or curriculum sequencing. These variations arise due to the difference in epistemology between mathematics and science; in other words, different ways of knowing and building knowledge. Lederman and Niess (1998) suggest that the nature of science and mathematics are fundamentally different. Although both share many features, the use and importance of empirical evidence and data is one area where they diverge. Both disciplines use logic to assess the importance of knowledge claims, but science must also refer to the external, empirical world whereas mathematics does not need to. As a result, in science there is an emphasis on context, on units and on variables that link measurements to the external world. This emphasis is not present in the mathematics curriculum to the same extent. However, as applied mathematics and statistics become more widely accepted elements of school mathematics, these differences may be diminishing.

Task 15.1 **Curriculum and assessments in mathematics**

Look at a copy of the mathematics curriculum and explore what is required for an aspect of mathematics that you will be teaching. Look at some examples of assessments in mathematics and the mark schemes. For younger pupils these might be end of year assessments in your placement school or external assessments (such as GCSEs in England and Wales) for older pupils. Choose a mathematical skill, for example graphing or solving equations.

■ What differences can you see between the types of tasks pupils might need to do in mathematics and in science?
■ What skills are required in science which are not taught in mathematics?
■ Write a bullet point list of the implications of what you have learnt for your teaching of science.
■ Discuss your findings with your tutor, mentor, and/or other beginning science teachers.

Teaching to do or teaching to think?

Many science lessons containing mathematics, particularly those which are about a particular equation, involve introducing the scientific idea, then the equation and then lots of practice using that equation. Or for graphing, the focus might be on producing the best possible graph rather than thinking carefully about what the graph means and how it relates to the data being graphed. Such lessons are usually (not always) teaching pupils to *do* but not necessarily teaching them how to *think*.

To teach pupils to solve problems then it is necessary that sometimes they get stuck, in other words to work through problems to which the solution is not immediately obvious. If pupils are given tasks which do not make them think then they may be exercises or practice of an idea, but they are not problems.

Getting stuck is not comfortable and pupils, particularly those who are used to finding work straightforward, can resist being given tasks which require them to work through something challenging. They may require lots of encouragement and I have sometimes told pupils that if they can do the work that I give them easily, then I am almost certainly not doing my job properly as they should be being required to think in my lessons. The skill as a teacher is to set up these sticky moments so that they support and develop pupils' self-efficacy (their belief that they can act to achieve specific goals) rather than leaving them feeling that what is asked is beyond them. In Vygotsky's (1962) terms this would mean that pupils are working in their *zone of proximal development*. It is usually more comfortable for pupils to be required to get stuck together so working in groups on challenging problems is a good way to support all pupils. To think mathematically requires that pupils can reason and justify their answers and group discussion helps to develop these skills too, if the task is set up carefully by the teacher.

One way to support pupils in learning to think, to reason, and to explore ideas mathematically is to use more open tasks, sometimes called rich tasks. Jo Boaler in her book *Mathematical Mindsets* (2022) suggests that there are six questions that if asked and acted upon in designing and setting a task will increase the power of that task to support mathematical thinking. They are:

1 Can you open the task to encourage multiple methods, pathways, and representations?

You might require pupils to show their thinking visually, with numbers, with fractions, or in other ways. For example, in this task from NRICH maths (2023a).

I mixed up some lemonade in two glasses.

The first glass had 200ml of lemon juice and 300ml of water.
The second glass had 100ml of lemon juice and 200ml of water.

Which mixture has the stronger tasting lemonade?
How do you know?
How might you use fractions to help you to work out which mixture is stronger?
How might you use ratios?
How about a graphical approach?
Given several similar problems:
Do you always use the same strategy?
Describe some occasions when one strategy might be more efficient than another.

---

2 Can you make it an enquiry task?

When pupils think their role is to come up with an idea rather than to reproduce a method then they will be far more engaged. For example, at the start of a lesson on pressure you might want pupils to recall how to calculate area. Rather than asking them to find the area of a 6 x 4 rectangle, ask them how many rectangles they can make with an area of 24. Squared paper can be helpful to encourage a visual approach.

Rather than giving pupils several examples of pressure to calculate, ask them to work out which exerts more pressure on the floor, an elephant or a person in high heels. Depending on how much time you wish the task to take you could give them approximate masses or leave them to make approximations and assumptions for themselves. The point is that they will be thinking mathematically about the problem, not just working through a list of calculations. Pupils should be prepared to work together, reason, and justify their answers.

3 Can you ask the problem before teaching the method?

This method works particularly well for teaching a standard method or formula, for example calculating surface area to volume ratio or calculating speed. For example, pupils can usually do speed, distance, time calculations in miles per hour before they have been taught the formal equation because they are so familiar with speed in their everyday lives. You could then ask them to try calculations using another unit - such as metres per second - and ask them to come up with the equation for themselves.

4 Can you add a visual (or physical) component?

This can be any other way of representing the problem that will help pupils to understand. For example, in calculating density, have samples with the same volume but different masses and ensure pupils take the time to feel the different densities before they rush to calculate them.

Adding colour can also be helpful. For example, when answering a calculation question pupils could be taught to circle or highlight each number in a different colour and colour code the equation they need to use to match.

Pupils can be encouraged to draw out any problems that they are struggling with, as using different representations can aid thinking and understanding.

5 Can you make it low floor, high ceiling?

A low floor task is one which is easily accessible, a high ceiling task is one which challenges all pupils. One way of making tasks higher ceiling it to ask pupils to write a question related to what they have been working on - they should try to make the question hard. You can either ask them to answer their own questions or to swap and answer each other's. They could also peer mark the questions they have set.

6 Can you add the requirement to convince and reason?

Reasoning is key to both mathematics and science. Being able to reason, argue, and justify your ideas promotes the learning of science (Evagorou & Osborne, 2010). Engaging in classroom talk about ideas in science is particularly beneficial to those who struggle with science learning (ibid). Research by Boaler (2022) in mathematics classrooms found that being required to reason diminished the achievement gap between pupils as it gave pupils access to others' understanding. Research (Grey & Tall, 1994) has shown that pupils with lower attainment in mathematics often use harder mathematics to get an answer than those with higher attainment. Reasoning and discussion help to open up different methods to those pupils. For the greatest impact, therefore, discussion groups should include pupils with a mix of prior attainment. This requirement can be added to any task. Pupils could be asked to convince themselves, a friend, or a sceptic.

The first two are relatively straightforward; the last much harder. This requirement has the added advantage of improving pupils' oracy alongside improving their mathematical skills.

Task 15.2 **Make a task richer**

Choose a lesson which you are going to teach which has a mathematical component. Use one or more of the previous questions to help you plan to make at least one of the tasks richer.

Reflect on the lesson and how the task worked with the pupils. Did they learn what you hoped?

Discuss your plan and the lesson outcomes with your mentor, tutor, and/or other beginning teachers.

What are the advantages and disadvantages to this type of task?

Using numbers and formulas

One of the key differences between science and mathematics is that in mathematics pupils are usually using numbers (or letters to represent numbers), whereas in science they must calculate with values, sometimes called quantity calculus. Values have a number and a unit with the units usually coming from measurement. This difference comes from the emphasis in science on the physical world and empirical data.

Redish (2017) argues that in physics (and this is true for all science) equations are linked to physical or biological systems and that this adds information about how they should be interpreted. In the simplest terms, it is not possible to add two numbers if they have different units; one cannot add length to mass. Scientists use physical knowledge when applying mathematics to physical systems, which might involve knowing the limits to the numbers which can be put into an equation. Redish (ibid) argues that physicists use physical knowledge in doing mathematics as much as they use mathematics in doing physics. Redish and Kuo argue that part of the 'acculturation of a physics pupil is learning to interpret the math physically, not to only focus on mathematical structure and manipulations' (2015, p. 567).

For example, in both mathematics and science equations of the type $a = b \times c$ are frequently used. For the speed equation (*distance = speed x time*) none of the values would be negative as speed, distance, and time are scalar quantities which always have positive values. There is no mathematical reason why a negative number could not be introduced into the equation, but there are reasons based on physical knowledge. However, in the similar equation for velocity (*velocity = displacement ÷ time*) displacement and velocity are vector quantities and can be negative. Pupils are expected to use their scientific knowledge alongside their mathematical skills in calculating speed and velocity. Speed calculations require little additional understanding and are often introduced to pupils at age 11. Velocity calculations, although they look very similar, require more scientific understanding and are therefore usually introduced later.

Pupils are usually comfortable with speed-distance-time calculations because they have interactive, personal experience with speed. They know that if they leave home later, they must walk to school faster. Or, if they wish to make a detour to pick up a friend, they either leave earlier or speed up. However, similar looking calculations, also of the type $a = b \times c$, can be much harder. For example, mole calculations using *number of moles = mass ÷ molar mass*. Pupils do not have direct experience of (and frequently do not understand) moles or molar mass. This makes calculating them far harder and, indeed, mole calculations are usually introduced much later on a pupil's journey through school science.

How to rearrange?

Many science teachers prefer to teach pupils how to get an answer to a calculation using formula triangles (Wong & Dillon, 2019). For example, a speed, distance, and time triangle. If you cover the variable you wish to calculate (for example speed), that leaves distance over time remaining. Hence, speed is distance divided by time.

There are problems with choosing to teach the calculations in science this way. First, using formula triangles is an example of teaching pupils a process rather than teaching them to think about what they are doing and what the quantities they are dealing with mean. It is a shortcut to getting an answer rather than thinking about the formula. Second, many pupils simply do not understand triangles, a problem which is exacerbated when the formula becomes more complex than $a = b \times c$. Furthermore, it is a wasted opportunity to apply their learning from mathematics and they can even undermine teaching in mathematics. Last, it is a wasted opportunity to think scientifically, to reason using mathematics about a scientific situation. Beyond this, in England, pupils will not receive credit in their GCSE external assessments for writing down a triangle

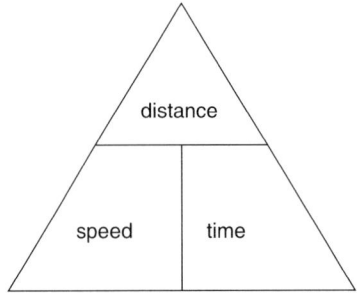

■ **Figure 15.1** Formula triangles can be used to help rearrange equations

if they are asked to give an equation or to rearrange it. They would get all the marks to a calculation using the triangle method correctly but would not necessarily get method marks if their calculation went awry.

There are reasons why science teachers choose to use triangles in their teaching. The curriculum includes many equations which must be mastered by pupils, 28 in physics alone for combined science GCSE in England (AQA, 2022). With this many equations to cover it is understandable that teachers look for a shortcut, but triangles are a shortcut that frequently reduces understanding. Many believe, often correctly, that pupils struggle to think mathematically about the formulas and that they feel more comfortable with triangles, although feeling comfortable can be a sign that learning is not taking place. Some science departments agree on a department-wide policy for tackling calculations with formulas and this can include the triangle method.

For many pupils, rearranging algebra is rather abstract. It can be helpful instead to model writing out the equation, writing the numbers you know into it first and then rearranging if necessary. You can also teach pupils to colour-code the values in the question and the terms in the equation (adding a visual component as discussed earlier) which can help them to match up which values should go where and reduce mistakes. If you can, it is worth talking to a mathematics teacher at your school about how they teach algebra so that you can use some of the same phrases. For example, 'what you do to one side, you have to do to the other' is often used as it promotes understanding, whereas 'cancel out' is not. It is also worth finding out when pupils are likely to learn to rearrange equations. In many schools, pupils at age 11 first meet rearranging equations in science rather than in mathematics. Having the expectation that pupils know how to rearrange is likely to lead to a frustrating lesson for both pupils and the teacher. Using those early lessons to teach for understanding and for long term success in using equations will ultimately support pupils in their learning in both science and mathematics and is therefore time well spent.

Task 15.3 **Talk to a mathematics teacher**

Talk to a mathematics colleague (either an experienced teacher or a student teacher) about how they teach an aspect of maths (volume, density, graphs, algebra, etc.) Is there an agreed approach in the mathematics team or does each teacher approach it differently? Find out when pupils are likely to meet this aspect of mathematics and if it is the same for all classes.

What are the implications for your teaching, particularly if pupils are in different groups in mathematics and science?

Discuss what you have learnt with your mentor. Are there any approaches to tasks agreed between the science and mathematics departments?

GRAPHS AND GRAPHING

Graphing in mathematics and in science

Graphs are important in both mathematics and science and thus developing pupils' graphicacy or graphical literacy is a key goal in both subject areas. Graphs are used in both disciplines to communicate information, to summarise results, and to display relationships.

In school mathematics, pupils are expected to be able to translate between algebraic and graphical representations. A graph is considered a representation of an exact relationship which can be described by a mathematical equation. For example: Draw the graph of $y = 0.8x$ for values of x from 0 to 6. (Axes are given; AQA mathematics.)

Due to the emphasis on the graph as an exact or idealised relationship, there is little treatment of errors – even when the mathematics is in a science-like context. For example, this question is from a mathematics paper:

Jim buys a plant of height 20 cm. The graph shows how the height of the plant changes during the next 4 days. Work out a formula for h in terms of n (AQA, 2020, p. 15).

There are no points actually plotted and it is highly doubtful that real data would all lie on the line. This is not a criticism but stems from the different emphases of the two curricula. The aim of this question is not for pupils to understand plant growth or the experimental method but to translate from a graph to an equation. Mathematical ideas may be better understood in a context pupils can relate to, but the goal here is the mathematics, not the science.

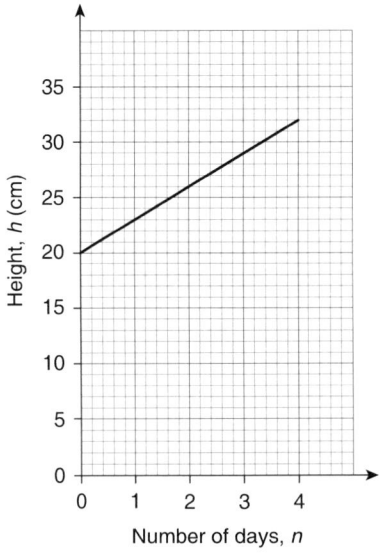

■ **Figure 15.2** A graph question from a mathematics paper in a scientific context

Source: AQA, 2020, p. 15

In contrast, understanding variation in data and errors are key in science. For example, pupils are given this graph of the data from a circuits experiment. They are asked to identify the anomalous result and explain why they have chosen it:

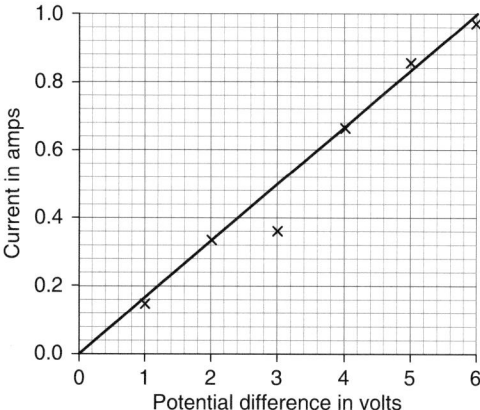

■ **Figure 15.3** A graph from a physics paper

Source: AQA, 2012, June 2012 CS physics paper 2 foundation (4451) Q2

As context is important in science, pupils are frequently expected to move between the graph and a real situation. For example, this graph shows results of a reaction taking place at 40°C. Pupils are asked to draw a curve on the graph for the results they might expect at 50°C.

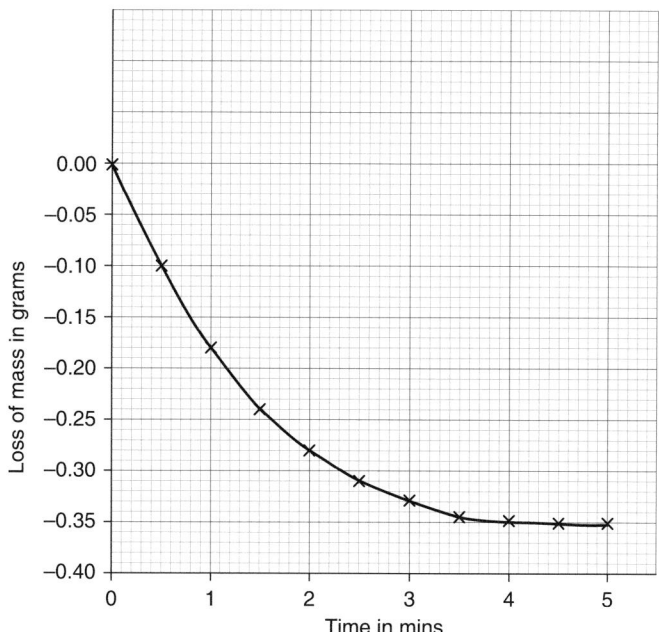

■ **Figure 15.4** A graph question from a chemistry paper

Source: (AQA, 2020) GCSE combined science, chemistry paper 2, higher tier, Q2

To answer the question, they must bring their scientific knowledge to the situation and understand that the reaction will be faster so the curve will be steeper. They must start the curve at the same point, and it must finish at -0.35:

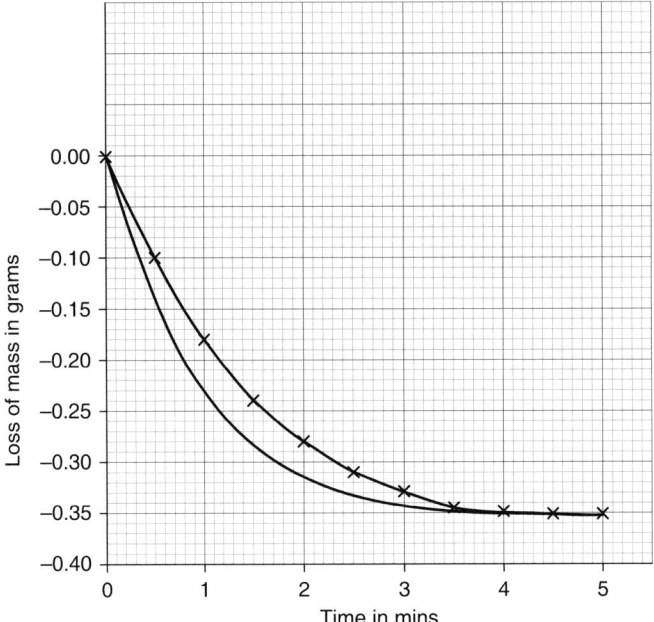

▪ **Figure 15.5** The answer to the graph question in Figure 15.4

Source: AQA, 2020. GCSE combined science, chemistry paper 2, higher tier, Q2

The key point here is that while graphs are used in both mathematics and science, the types of tasks pupils are typically asked to carry out vary. Although there are some aspects which are common to both (such as plotting points or calculating a gradient) there are also many differences. It is not sufficient to notice that pupils find graphing difficult and to do nothing about it (as Goldsworthy et al. (1999) found); the teaching of the graphing skills required for science must be planned and taught explicitly. Furthermore, pupils often meet graphing in science before they meet it in mathematics and so science teachers need to be very careful in what they expect of pupils.

Graphing skills

Leinhardt et al. (1990) break down the skills that pupils require into two broad categories, interpretation and construction, although they are not mutually exclusive.

Construction

Construction might include plotting points or drawing a line of best fit and these may overlap with the mathematics curriculum. It might also involve making predictions, for example extrapolating from an incomplete graph or beyond the data plotted. In science, this frequently involves understanding the context and also understanding what the representational graph should look like. The reaction rates question above is an example of this type of construction.

Scaling tasks are those relating to axes, scales, and units. In virtually all graphs in science the units are important as they are what relate the graph to the real-life situation. Often the units are different on each axis and thus the scale may also be different. Learning to choose and construct a sensible scale is a key part of learning to graph in science.

Many pupils will be overwhelmed if they are expected to learn all these skills at once. Careful consideration of the learning objectives for any task will allow you as a teacher to decide whether for any given lesson it is more appropriate for at least some pupils to be given partially completed graphs to allow them to focus on only one or two specific skills.

Task 15.4 **Realistic timings**

Many beginning teachers find it difficult to set realistic timings for activities, and graphing is a task which often takes longer than is planned for. Time yourself drawing a graph of some data. How long did it take? How long to find a pencil, rubber, and ruler? How long do you think it would take a 15-year-old to do the same? Or an 11-year-old?

Observe a science lesson where pupils are graphing. Pick one or two pupils and time how long it takes them to do each part of the task they have been set (e.g. choose axes, select the scale, plot the points, join them up).

Next time you plan a lesson involving graphing think carefully about what exactly you would like pupils to do and whether you have sufficient time or need to reduce the number of graphing tasks you are expecting pupils to complete.

Interpretation

In addition to being able to construct graphs, pupils are expected to understand and interpret them. Interpretation often involves moving between the graph and the situation it represents. For example, to interpret a graph of enzyme action may also require understanding the effect of temperature, such that an increase in temperature may not lead to a consistent increase in reaction rate. Both an understanding of the graph and an understanding of the scientific situation are required to make sense of the data. This skill is less often required in mathematics education and thus science teachers need to prioritise teaching pupils to understand and interpret graphs of real-world situations.

Graphing tasks

These tasks all aim to develop pupils' conceptual understanding of graphs rather than the process of drawing them. The requirement to convince and reason can be added to any of these tasks.

1 Walking graphs

Draw axes onto the floor. This could be done in chalk or with tape, inside or outside. Have large labels for the axes and encourage pupils to walk the pattern they might expect to see. This could be done for predictions, data collected, or a situation you discuss.

2 What could this graph represent?

Project a graph with the labels for the axes (and possibly the scales) covered up. Ask pupils what it could be a graph of, encouraging outlandish answers at the start. As you reveal more information the possibilities reduce. This works well as a starter activity.

3 Fill me up?

Show pupils some science equipment. Imagine you put them under a steady stream of water. Sketch the graph of height of water level against volume as the containers fill up (NRICH, 2023b).

4 What does this graph tell us?

Project any graph. Ask pupils to discuss what they can learn from the graph. If appropriate, they could also discuss what would improve the graph (clearer axes, better labelling, etc.) This is another good starter activity and works particularly well with graphs from recent news items or environmental data.

Task 15.5 **Observe a mathematics lesson**

Observe a mathematics lesson. It would be most useful to observe one on a topic related to science such as graphs, rearranging formulas, or unit conversion.

In advance of the lesson, consider what expectations you would have of pupils in science for this idea; think about how you would introduce the concept to a class and compare this to what you see.

If you can, talk to the teacher after the lesson and seek to understand the rationale to the approach they have taken.

How might you use what you have learnt when planning your next science lesson which requires pupils to learn or use this skill?

Many schools offer early career teachers the opportunity to observe experienced colleagues and you could choose to observe a mathematics colleague at this point too.

SUMMARY AND KEY POINTS

Many authors suggest that the way to solve the difficulties that pupils often have with using mathematics in science is to collaborate with the mathematics department (Wong & Dillon, 2019). While on the surface this is sound advice, in reality it rarely occurs and is challenging to sustain. Although collaboration is a worthwhile aim, it is not a simple solution to pupils' difficulties.

Instead, teaching the mathematics required for science is part of the role of the science teacher. The types of tasks that are required of pupils are not the same as those prioritised in mathematics lessons and so expecting that pupils will simply arrive with the mathematics skills they need for science is not appropriate. The teaching of the mathematics in science requires the same careful planning as any other aspect of the curriculum. Aiming to support pupils to think mathematically in science rather than to simply follow procedures may contribute to helping them develop as mathematicians; it will also allow pupils to develop a better understanding of science and is thus time well spent.

FURTHER RESOURCES

Boaler, J. (2022). *Mathematical mindsets* (2nd ed.). Jossey-Bass.

An inspiring book which encourages teachers to support pupils to think about mathematical ideas. Although it is mainly written to mathematics teachers, there are several ideas which could be used effectively in science lessons.

Boohan, R. (2016). *The language of mathematics in science*. https://www.ase.org.uk/mathsinscience

Mathematics and science educators collaborated in this publication which aims to help teachers understand the differences in language between mathematics and science.

Mason, J., Burton, L., & Stacey, K. (2010). *Thinking mathematically* (2nd ed.). Pearson Education Limited.

The aim of this book is to develop pupils' mathematical thinking in any context. The later chapters are written to teachers with ideas about developing mathematical thinking in others.

NRICH maths supporting SET. https://nrich.maths.org/9517

This website from Cambridge University includes many rich mathematical tasks including some in science contexts.

Wong, V. (2017). Variation in graphing practices between mathematics and science: Implications for science teaching. *School Science Review, 98*(365), 109-115.

An exploration of how and why graphing varies between school mathematics and science and the implications of those differences for science teachers.

16 FORMATIVE ASSESSMENT IN SCIENCE

Ed Walsh

INTRODUCTION

Formative assessment, also referred to as 'Assessment for Learning', is based on the use of assessment to improve teaching and learning (see e.g. Black et al., 2004). The teacher and pupils should be able to use evidence from assessment to guide next steps in teaching and learning. Assessment should not just be reporting on progress in learning but actively supporting its development. There are several aspects to making a success of this. You will need to consider activities that are effective as instruments of assessment. You will need to think how to integrate such activities – effective lessons have assessment built in from the outset, rather than bolted on as an additional feature. You will need to respond to the evidence yielded by the assessment. There is little point to formative assessment if later parts of the lesson proceed as originally planned, irrespective of how pupils are responding. Whether formative assessment shows that pupils don't understand something or already understand elements well before you begin a topic, it enables you to be flexible and responsive, teaching a key concept again in a different way or moving on more quickly to build on pupils' pre-existing knowledge. Insights gained from formative assessment may also change your approach on a wider basis to teach other groups differently. Another crucial outcome of formative assessment is that pupils themselves become more effective learners. They should be able to draw conclusions from the assessment evidence, both in terms of what they have learned and how they learn, which will foster better learning behaviours/a more self-regulated approach to learning (Quigley et al., 2018).

We begin by exploring how to plan for assessment in science before discussing how to use insights to inform teaching and finally taking a look at the role of formative assessment in pupil self-regulation of their own learning.

DOI: 10.4324/9781003110187-17

OBJECTIVES

At the end of this chapter you should be able to:

■ Understand the form and function of a range of formative assessment strategies and be able to incorporate these into lesson plans, including consideration of lesson design, questioning, feedback, and self and peer assessment.

■ Use these strategies effectively with pupils to elicit evidence about learning that has taken place.

■ Act on this evidence to make teaching more effective.

■ Understand how to use formative assessment to develop pupils' own capacity as learners.

PLANNING FOR FORMATIVE ASSESSMENT

Formative assessment practices need to be purposeful, to collect evidence about pupils' learning, and to enable pupils and teachers to act on that evidence. We will begin this chapter by focusing on lesson design: how to plan for formative assessment. The key areas we focus on here are:

1 The relationship among objectives, outcomes, activities, and assessment, drawing on and extending the thinking in Chapter 8 (planning for progression)
2 The use of questions and questioning

Outcomes, objectives, and activities

Formative assessment is important in teaching and learning in all contexts, and often schools will have expectations which are applied across the institution, meaning that pupils should have an idea of how to improve in whatever they are studying and be able to enact this within each subject discipline. However, there are some aspects in which formative assessment practice needs to be customised to the subject being studied. For example, in science there is a need to balance the roles of conceptual knowledge, procedural knowledge, and manipulative skills. This will inform the framing of objectives and outcomes.

As discussed in Chapter 8, when planning it is crucial that you are clear about what you want pupils to know and be able to do as a result of participating in the lesson. In other words, framing your lesson objectives and outcomes at the start of your planning. One reason it is crucially important to get these right is that they should encapsulate what the lesson is about and provide a yardstick by which the progress pupils are making can be identified. In many schools there is an expectation that learning outcomes will be not only planned but shared with pupils. This can be helpful but it is possible for teachers to lose sight of their function. If they are there because they are a requirement of the institution rather than because they are driving the design and delivery of the lesson, they run the risk of becoming tokenistic.

With a view to formative assessment, there is a strong case for framing outcomes as actions: writing them to indicate what pupils should be able to do (and which they couldn't do or, at least, couldn't do as well at the start) can help the teacher think about outcomes in terms of evidence. What will you be able to see, as a teacher, if pupils have succeeded? It might be a piece of writing, it could be a drawing annotated with key features, it might be an explanation given orally or it might, for example, be a focused image of a plant cell under a microscope. Planning outcomes in this way makes designing or selecting formative assessment strategies easier, because we know what we're looking for.

For example, a set of learning outcomes might look rather like this:

- ▪ To be able to recall the relationship among pressure, force, and area
- ▪ To calculate the pressure applied from the force and the area
- ▪ To apply this to a range of practical contexts.

Outcomes can be progressively more challenging and, indeed, some schools require this in the planning and delivery of lessons. One of the structures you may see used is that of 'all, most, some', referring to the level of challenge of an objective and the extent to which all pupils in the class might be expected to achieve it. However, it's important to think through the implications of this. Lessons should be designed and delivered so that all pupils can access the learning; rather than having lower attaining pupils never being able to access more challenging ideas, the focus should be upon the amount of support that is needed to achieve this. This is an example of a growth mindset in action, seeing pupil capacity as not being fixed but capable of being developed.

The next stage is to frame objectives, which should have a strong relationship with the outcomes. The objectives should indicate what pupils are learning to do during that lesson. Again, getting these right is important – it's what they are learning to do, not what they are learning about. They are learning objectives, not a crib sheet of content. One of the criticisms of the role of learning objectives is that they may 'give the game away'. They tell the pupils what is to be learned and therefore are the enemy of them discovering things for themselves. There is a real role for pupils in finding things out and if objectives are undermining this then they are doing us a disservice. However, this is down to the way they are being framed. For example, if we are going to get pupils to learn how to identify common gases using simple practical tests, then our outcome might be 'Pupils will be able to identify hydrogen, oxygen, and carbon dioxide using tests'. The corresponding objective could then be 'We are learning how to identify some common gases' (and *not* 'We are learning that hydrogen ignites with a squeaky pop, etc'.).

The learning activities can then be thought of as the bridge between the objectives and the outcomes. If pupils do these activities, then it should support progress towards achieving the outcomes. As Dylan Wiliam says in 'Embedded Formative Assessment':

> I often ask teachers 'What are your learning intentions for this period?' Many times, teachers respond by saying things like, 'I'm going to have the pupils . . .' and then specify an activity. When I follow up by asking what the teacher expects the pupils to learn as a result of the activity, I am often met with a blank stare, as if the question is meaningless or trivial. This is why good teaching is so extraordinarily difficult. It is relatively easy to think up cool stuff for pupils to do in classrooms, but the problem with such an activity based approach is that too often it is not clear what the pupils are going to learn.
>
> (Wiliam, 2011, p. 61)

Task 16.1 **Developing meaningful objectives**

- Select a lesson that you are planning, either to support or to teach
- Develop a set of learning outcomes that clearly identify what pupils should be able to do by the end of the lesson. Frame them as behaviours. Try to make these progressively more challenging and try to base them as much as possible on the curriculum plans for the course, such as the exam specification
- Now write objectives to share with pupils at the start of the lesson
- Outline the kind of activities that might now support progress towards the outcomes.

Questions

Good teachers ask lots of questions, using them for a variety of functions. Finding out what pupils know and understand is an obvious purpose, but it is only one of many. Sometimes teachers tend to make extensive use of closed questions which can be devised 'on the hoof', answered quickly and with a brief response. These are fine but may be of limited use in finding out the full range of what pupils can really do. For example, if a stated outcome of a lesson is that pupils should be able to apply ideas of energy transfer to melting ice then it's appropriate to ask questions that need longer explanations, such as 'What happens to the heat energy at the melting point?' This might seem obvious, but there is an associated skill set that is required. If pupils are used to quickfire closed questions, then training them to develop a longer response will need strategies and practice. One such approach is to ask pupils to 'think – pair – share' a question, in which they first think, on their own, of a response, then discuss it with a peer and then share it with the class. Another is to ask pupils to work in twos or threes to draft a sentence, which you can then call upon particular pupils to share. The purpose in each case is to encourage reflection and improvement in the quality of responses.

This is not an argument against closed questions but rather a suggestion that they should be just part of the diet. What good teachers will often do is to start off with some closed questions and then move into more open questions or ones that need a longer response. For example:

1 In which ways is electricity is generated in this country?
2 Which of these ways use renewable sources?
3 Why might a country want to move towards a greater use of renewable sources?
4 What do you consider to be the potential of wind farms as a supply of electricity?

The earlier questions here are clearly more accessible and there could be a reasonable expectation that all pupils could contribute ideas. The later ones are more open ended and are better answered with longer responses. This is known as 'ramping' and can be applied to questions whether they are to be done individually or in groups.

It's also important, especially when considering the role of formative assessment in lessons, to manage who answers a question. An increasing number of teachers don't ask pupils to put their hands up to indicate willingness to answer a question but select who will respond. This is important as the teacher should be aware of overall level of understanding. If the teacher accepts a good response from one of the few pupils who can answer well and then presses on, they may end up trying to build progress on foundations of sand. However we need to be aware that there are multiple purposes to the asking of questions; teachers will often use them as a means of control. There is an important difference between a teacher eliciting responses from a range of pupils to see if the group has understood and the targeting of questions at apparently disengaged pupils who may then feel under duress and even become less invested in the learning.

Teachers often refer to the 'climate for learning' in a lesson and the importance of keeping it positive. If pupils don't understand something, despite making a reasonable effort, then it's not their fault. An effective teacher will want to find out if effective learning hasn't taken place and will want to do something about it. It's also worthwhile considering various ways of gathering responses and sometimes using ones that avoid asking pupils to verbalise ideas in front of a class. This might, for example, involve pupils displaying responses on dry wipe boards (sometimes referred to as 'show me' boards) or using apps such as Padlet or Socrative.

A laudable aim regarding climate for learning is to make it high challenge but low risk. Pupils will be challenged to think in different ways and make sense of ideas and contexts they have not encountered previously but they know (both through your assertion and their experience of your teaching) that this doesn't mean that they will be made to feel uncomfortable.

Another aspect of using questions is to consider how to manage ones that pupils raise. A teacher who is good at engaging pupils and stimulating ideas is likely to be rewarded by pupils asking questions. Sometimes a teacher's response is to answer those themselves; sometimes this is appropriate but on other occasions a skilful practitioner will use this as the basis for inviting ideas from other pupils. What is important from all use of questions though is for the teacher to consider what they learned from the pupils' responses.

Task 16.2 **Making questioning more effective**

- Consider a particular lesson that you are planning, either to teach or to support
- Devise a set of questions that could be used with pupils; these should be ramped so that the earlier ones are more accessible and the later ones more challenging
- Suggest a strategy to use to maximise participation in the answering of these questions.

Hinge questions

A hinge question is a question that can be quickly and effectively posed and responded to during a lesson that will indicate whether it is appropriate to continue with the lesson as planned or whether the teacher needs to find a way of revisiting a particular concept or skill before further progress can be made.

The idea of checking progress and identifying gaps in knowledge is nothing new; teachers sometimes run a short test part way through a topic to see how well pupils are doing. The hinge question is a rather different device though; it focuses on a learning point that has a crucial role in the topic. It should provide an immediate response; these are often framed as multiple choice questions. For example, in a lesson on centripetal force, a teacher might present pupils with a question like the one shown in Figure 16.1.

The hinge question associated with Figure 16.1 might be: A ball bearing is fired into a curved tube in the direction shown. The view is from the top down, looking onto the bench top. When the ball reaches the end of the tube, which of the four arrows best describes its onward journey?

The teacher would use this to elicit responses from pupils. This might be done, for example, by getting pupils to write responses on dry wipe boards (sometimes referred to as 'show me boards') and display their answers. The teacher then makes a call as to whether to proceed to the next stage in the plan or to revisit the ideas from the first part.

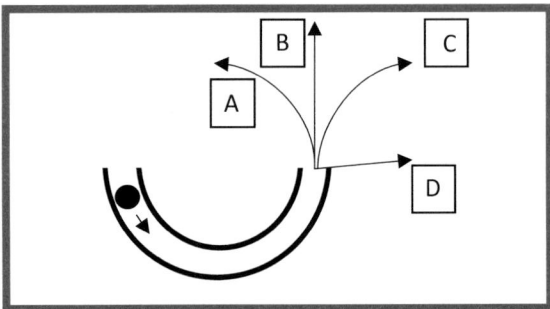

■ **Figure 16.1** Aerial view showing ball travelling along curved tube and then emerging. Four possible routes have been suggested. The teacher could use a question like this to gauge student responses and therefore levels of understanding

A couple of points are worth making around this. The first is that effective teachers don't revisit the earlier part of the lesson by just repeating what has already been said (after all, that clearly wasn't entirely effective). It needs to be a different way of explaining the idea and that's something that a less experienced teacher will need to think through in advance.

The second thing is that no matter how pupils are grouped, you are likely to get a range of responses to the hinge question. It would be naive to imagine that either all pupils will get it or that no pupils will. You therefore need to think through how to respond to a situation in which some pupils come up with a positive response and others don't. It might be that if there is a smaller proportion of pupils giving positive responses to the hinge question that these pupils are set a task that moves them on to the next stage whilst you revisit the earlier idea with the rest of the class. It might be that if table groups are designed to be mixed prior attainment groups that pupils who did understand the idea are then asked to take other pupils through other examples.

Teachers who devote time and effort to developing the role of questioning and making it more effective have often found it useful to follow these three suggestions from 'Working inside the black box' (Black et al., 2004):

■   'More time has to be spent on framing questions that are worth asking, i.e. questions which explore issues that are critical to the development of pupils' understanding.

■   Wait time has to be increased to several seconds in order to give pupils time to think and everyone should be expected to have an answer and to contribute to the discussion. Then all answers, right or wrong, can be used to develop understanding. The aim is thoughtful improvement rather than getting it right first time.

■   Follow up activities have to be rich, in that they provide opportunities to ensure that meaningful interventions that extend the pupils' understanding can take place'.

One of the (many) tensions a teacher has to resolve is the best use of limited time. Sometimes it may seem that responsive teaching is a great idea but that the time to revisit ideas doesn't exist. You've barely got time to cover what's in the course, never mind go back over ideas that pupils didn't get the first time around. In 'Responsive Teaching' Harry Fletcher-Wood suggests three beliefs about teaching:

1 'There is more that we would like to do for pupils than there will ever be time for; we cannot complete every desirable task.
2 Not all tasks are equally valuable; completing a worthwhile task may prevent us from doing something even more important.
3 We cannot solve these challenges by creating more time, being more efficient or juggling goals, because the possible tasks will remain unlimited'.

In other words you will have to repeatedly make judgment calls on two key things: what are the most important ideas for pupils to master and what are the most powerful ways of addressing them?

Task 16.3 **Developing the effective use of hinge questions**

■ Consider a lesson plan for a topic that you have coming up soon
■ Think about ideas that you will be introducing in the earlier part of the lesson and devise a hinge question that would reveal whether pupils have grasped the idea
■ Decide on the kind of response that would make you feel confident about proceeding with the lesson
■ Plan for strategies suitable for responding to less positive responses to the hinge question, both in terms of alternative teaching and also pupil management.

Feedback

Teachers provide feedback to pupils in a variety of ways. Good teachers will take many opportunities to indicate to pupils how they are doing and will also ensure that many of these are positive. These teachers understand that there are different kinds of feedback (see later explanation) and that some are more effective than others; general comments of a bland nature tend to have little impact. Teachers respond to pupils' verbal contributions in lessons, they make comments on written work, they may have one to one conversations with pupils, and they communicate with parents and carers. Whereas these don't all have quite the same function, it can be expected that they all lead towards more effective learning.

Research shows that feedback has the potential to be powerful. Professor John Hattie's Visible Learning meta study identifies the effect size of feedback as being 0.7, which places it in the top category of 'potential to considerably accelerate pupil achievement' (Hattie, 2012). However, this doesn't mean that *any* feedback will have such a positive effect. There are ways to make it more effective.

In 'Responsive Teaching' (2018) Fletcher-Wood identifies four different levels of feedback, ranging from specific to general. It may be that the feedback is focused on the specific task and guiding the student as to how to do a better job of that. This is easy to do, though time consuming, but it's important to consider how easily the student can then apply this to other aspects of their work. It may be that the feedback is more focused upon deepening understanding of the subject, considering, for example, how a practical procedure can be usefully structured. Feedback may also focus on self-regulation or even self-evaluation, encouraging the student to consider how effective they are at learning.

It's common in science lessons to see task-related feedback used quite a lot. This is appropriate in that pupils sometimes need specific guidance, but its limitation is that it's difficult for them to transfer the feedback to another task. Subject-related and self-regulation feedback have the potential to be more powerful, though this may be related to the wider dialogue in the lesson. Some teachers are better than others at explicitly exploring with pupils what it means to be an effective scientist. If this is part of their classroom experience then feedback along these lines is more likely to be effective. However, all four types have their place.

As so often happens in teaching, it is the detailed execution as well as the underpinning concept that will aid success. Black and Harrison (2004) ask readers of 'Science inside the black box' to compare these two comments:

- Add notes on seed dispersal.
- Can you suggest how the plant might disperse its seeds? Could this be an advantage?

Whereas the first comment gives the pupil a task to perform to improve his or her work, the second comment initiates thinking immediately. This enables the learner to discuss his or her thoughts either with the teacher or a peer and its questioning nature encourages the pupil to initiate improvement. The first comment simply describes a deficit in the piece of work.

Marking can be quite onerous so it's important to make the process efficient as well as effective. One example of this is to use a code to indicate improvement points. You could then either supply pupils with a generic sheet showing the corrections by numbered point or cover this the next lesson, again by point. This depersonalises the corrections (and also makes marking quicker).

Another technique increasingly used is that of live marking. In 'Making every science lesson count', Shaun Allison explains and justifies this,

ask them to respond to (a question) and then move away. Come back in a few minutes to check that they have responded. This is a very powerful form of feedback because it's in the context of the work they are doing, it makes them think and it requires them to do something straight away.

(Allison, 2017, p. 101)

It's also worthwhile saying that there is an important link between feedback on assignments and the setting of those assignments. As a general principle a teacher should be able to praise something that a pupil has done well and also to point out next steps in learning. Therefore the assignment should be selected so that all the pupils in a group should be able to attempt it, gain some success, and have this recognised. This is particularly important for lower-attaining pupils so that their confidence is developed. Similarly, assignments should always be set, as far as possible, so that higher attaining pupils can be provided with feedback on how to improve further.

Task 16.4 **Making written feedback effective**

- Consider a piece of written work that it would be appropriate to provide written feedback on
- Reflect on the different types of feedback discussed above (task-specific, subject, self-regulation, self-evaluation) and decide on which of these you could usefully use
- If there are types of feedback that wouldn't be applicable to that sort of assignment then consider situations in which they would be appropriate.

Self and peer assessment

A fundamental feature of effective practice is for pupils to know what is expected of them. There needs to be a clear indication of the success criteria that are being applied. For example, if pupils were being asked to develop a practical procedure, then the criteria might be:

- Steps set out in a logical order
- Clear instructions that are unambiguous and easy to follow
- Procedure which, if followed, would yield valid evidence.

These should be clearly conveyed to the pupils (actually at the point at which the task is set rather than waiting until the assessment point) and used by them (in peer or self-assessment), or by their teacher, to structure the feedback. There are various ways of setting this out but one that is in common use is 'what went well' and 'even better if' (often shortened to www and EBI). It is then appropriate of course for pupils to be asked to implement those improvements; if this doesn't happen then they may take the task less seriously next time around.

Peer assessment requires pupils to work effectively in small groups and needs them to be good at functional speaking and listening. This kind of talk needs to be actively promoted in the classroom. A good source of perceptive research and suggestions for developing practice is the work of Neil Mercer (see e.g. Mercer et al., 2004). For example, this set of rules for group talk was adapted by the EEF from Mercer's work):

> All group members must contribute; no one member should say too much or too little. Team members should encourage those who are saying less;
>
> Every contribution should be treated with respect, listened to thoughtfully, and allowed to finish;
>
> Each group much achieve consensus by the end of the activity, and you may need to resolve differences;
>
> Every suggestion a member makes has to be justified – say what you think and why you think it.

(EEF, 2018, p. 16)

Task 16.5 **Getting pupils to devise feedback**

- Find an opportunity to get pupils in one of your groups to either self or peer assess (or both) a piece of work they have done
- In preparation for this develop a set of success criteria and share this with the pupils
- Get pupils to do the assessment and then look at some of the feedback generated
- Evaluate this feedback and judge the extent to which it will, if responded to, give rise to an improvement in the quality of the work.

DEVELOPING THE CAPACITY OF PUPILS AS LEARNERS

Lessons tend to be highly structured situations. Teachers expend time and effort on devising plans with a variety of types of learning activities. This is partly because we want pupils to learn and understand things that they do not yet have a grasp of. However, we also want them to become effective learners, capable of self-regulation and being able to work independently. Learning how to learn is a key aspect of the enterprise of education (see Chapter 2 for more information about self-regulation and metacognition).

Formative assessment has a key role to play in this. Strategies should enhance the capacity of pupils to understand how they are progressing both within the specific aspects of a subject but also as learners more generally. They should be able to review their own progress, develop a sense of what they need to focus on next, and know how to address this. For some pupils this may be a long journey, but these skills will need to support them through education in a variety of settings and through life. Pupils may be used to such structures from their wider experience of schooling. However, they may not associate it with learning in science, so be prepared to spend time establishing and reinforcing this.

Feedback has a role in developing pupils as learners as well. Effective feedback is two-way and if the climate for learning is positive and well established then pupils will be able to give feedback to the teacher on what makes learning effective. Imagine the situation in which you are reviewing the structure of the human heart with a class. There are various ways that you could do this but it might be worthwhile asking the pupils what they feel would be the most effective way. You will still need to make a call on the approach to use and may choose to be guided by a particular suggestion or a broader consensus; an important aspect though is that you are developing the role of reflecting on how to learn well. This isn't, by the way, an abrogation of your professional responsibility to manage learning; the final decision on choice of approach remains yours, taking their ideas into account.

Task 16.6 **Developing pupils as critical learners**

Think about a class that you have worked with for a little while and consider how they are developing as learners. In particular you could get them to consider and talk to you about:

- What helps them learn more effectively.
- If they were revising a topic, whether they are aware of and have evaluated various strategies.
- How good they think they would be at learning something independently.

SUMMARY AND KEY POINTS

In this chapter we have explored a range of ideas about formative assessment, considered various aspects of classroom practice informed by it, and looked at some practical strategies.

- There are a number of aspects of lesson design which need to be planned and deployed to elicit useful evidence about whether effective learning has taken place. These include outcomes, objectives, and activities. The outcomes and objectives need to focus on the most crucial aspects of learning; the activities need to be engaging but also to provide learners with the best possible chance of achieving the outcomes
- There should be a range of types of questions used, and they should be used in different ways so as to promote deeper learning, maximise participation, and elicit useful insights
- Feedback can be powerful but to have the maximum impact it needs to be positive, focused on the central ideas, and lead to improvements
- Self and peer assessment can be used to give learners a clearer idea about what they are doing well and how they can focus their efforts to improve.

Formative assessment is predicated upon the teacher not only gathering and interpreting evidence of learning but also being prepared to respond to it and being able to structure learning in a different way so that pupils can succeed. It is also likely to be more successful if the teacher has established a positive climate for learning in their lessons.

Finally, formative assessment aims to not only enable to pupils to succeed in mastering a particular area of the taught curriculum but also to support their development as learners in a more general way, enabling them to work more independently and with a greater degree of self-regulation.

FURTHER RESOURCES

Black, P., & Harrison, C. (2004). *Science inside the black box*. GL Assessment, and Black, P. J. et al. (2004). *Working inside the black box*. King's College.

These two booklets followed the impact and success of the seminal 'Inside the black box' and explore the questions raised in more detail. Heavily based on empirical research and clear on recommendations for action, these are still essential reading.

EEF. (2018). *Improving secondary science*. https://educationendowmentfoundation.org. uk/tools/guidance-reports/improving-secondary-science/

This report has had a profound and significant impact both within and beyond secondary science departments in the UK. It aims to address the significant attainment gap affecting disadvantaged pupils and does so by identifying seven practical evidence-based recommendations. These are well argued, relevant, supported by research findings, and include clear injunctions for action.

Fletcher-Wood, H. (2018). *Responsive teaching*. Routledge.

This book starts by posing seven fundamental problems that face every teacher trying to get formative assessment to be effective. It draws on a wide range of case studies and also upon the developing role of cognitive science in making practice effective.

STEM Learning *Assessment for Learning in Science* resources. https://www.stem.org.uk/ resources/community/collection/49254/assessment-learning-science.

This is a collection of resources which are designed and selected to be used by practising teachers to support various aspects of practice in formative assessment.

Wiliam, D. (2011). *Embedded formative assessment*. Solution Tree.

This book is a useful account from one of original 'Inside the black box' authors that explores the potential impact of all aspects of formative assessment and includes examples of practical techniques.

SUMMATIVE ASSESSMENT IN SCIENCE
Ed Walsh

INTRODUCTION

Summative assessment has a significant impact upon the design and delivery of the science curriculum in secondary schools. The grades awarded at the end of a course matter to individual students both as a perceived validation of their status and effectiveness as learners and as a gatekeeper to subsequent stages in education. They also matter to institutions as they may be used as performance indicators. Schools that secure better grades usually attract and retain students, which secures them more funding, so understanding the form and nature of this assessment is important. Summative assessment may, depending upon its form and function, only recognise certain aspects of what is valuable in science as a subject. In a high-stakes environment other features may become marginalised.

Awarding organisations are required to satisfy a number of objectives in the development and deployment of summative assessment. It is important to understand what is being aimed for.

- Summative assessments are expected to be reliable. Students exhibiting a similar level of performance as defined by the course should receive similar grades and this needs to apply not only within a cohort but over successive cohorts. Reliability is achieved by measures such as the design framework for exam papers, standardisation – in which different markers practise marking items before they look at student responses – and moderation, in which the assessment of the work of different candidates is compared
- Summative assessments should be valid in that they should assess what is of value in an area of study. We expect students in science to, for example, have understood a range of concepts, to be able to apply them to a variety of processes, and to have developed appropriate skills. Students who get good grades should, overall, be good at all of these things. However, some

DOI: 10.4324/9781003110187-18

things that are valued in science are not as easy to assess in an exam as others

■   It is also often the case that summative assessments are required to discriminate between students. In other words, they should produce a range of grades. Unlike a driving test in which the outcome is a pass or fail, there is the need to differentiate different levels of performance.

Exam papers are technical instruments, designed in a particular way to do a certain job. We will explore what that job is, suggest how well they can do it, look at some ways in which students can be supported to perform well in exams, and reflect on what conventional forms of summative assessment are less effective at.

OBJECTIVES

At the end of this chapter you will be able to:

■   Describe the structure and purpose of summative assessment
■   Reflect on the implications of summative assessment for teaching and learning
■   Understand the limitations of summative assessment as a tool for the design of teaching and learning and the importance of teachers having a view of the aims of science education which is broader and deeper than that of the requirements of examiners.

SUMMATIVE ASSESSMENT

There are two parts to summative assessment. One is the syllabus or specification. This indicates the content of the course being offered and indicates to teachers what they should cover in their teaching, including scientific concepts, applications, and the requirements for scientific enquiry. It may indicate other aspects such as the role of maths skills. Furthermore, students will be expected to use the concepts and ideas they have learned, such as being able to apply them to novel contexts. The other part is the exam itself, which contains a range of types of question. Successive series of examinations will be different but they are required to have the same relationship with the specification.

Understanding the nature of summative assessment is important and useful for three reasons. First, if you understand how papers are constructed and how they are designed to achieve their intended purposes then you will be better able to prepare the pupils you teach to succeed. Part of being a teacher is the preparation of pupils for assessment so they are able to demonstrate what they know and can do. This includes covering the content but also relates to reflecting the nature of assessment objectives in lesson design and getting students to be familiar with command words.

Second, understanding summative assessment means that you can make effective use of the evidence it generates. Examiners aren't doing the same thing as teachers; they are required to award an overall grade that reflects performance in assessments whereas teachers need to not only recognise attainment but also plan next steps in learning. If a student part-way through a course gets 51 per cent on an exam and their target was 65 per cent, that information indicates underperformance; however it gives neither the teacher nor the pupil any indication as to how to address the gap. Knowing how the pupil got the marks they did and also where they missed opportunities enables us to target support.

Third, it informs us as to how we can design our own assessments if they are ones that are expected to predict exam performance. An end of year exam, for example, that is heavily based on knowledge and understanding, won't give a valid indication as to how pupils will do on exams where many of the marks go on application and interpretation and will mislead pupils as to how they should prepare themselves.

However, it shouldn't be assumed that summative assessment only relates to external assessment or to teachers aping the characteristics of external assessment to predict pupil performance in such exams. A teacher or school may decide that it is useful to assess what pupils know, understand, and can do by the end of a section of work irrespective of external assessment, for example in Key Stage 3.

Summative assessment and the nature of science

We need to consider the relationship of three factors. What this diagram (Figure 17.1) does is suggest the role of each of these.

The first factor is the nature of science as a discipline. This might well include the role of a range of key concepts such as the particulate model of matter, the gene theory of inheritance, and the model of energy transfer; we might also think it important that these can be applied to novel contexts. This is more than just a list of concepts though; we might also consider it important to recognise that, for example, scientists will argue over the interpretation of data gathered from an emerging area of science and defend a conclusion. We also might want to recognise such features as scientific knowledge being provisional in nature.

The second factor is what can be included in a school science curriculum. This might include, for example, introduction to and development of understanding in a range of ideas, exploratory work by students into phenomena, and the use of various types of practical work. This is likely to be influenced by a number of other factors, including qualifications being aimed for, the teacher's understanding of the nature of the subject, and the ethos of the school.

The third factor is what can actually be assessed using the tools available in the system we are working in. For example, it's fairly straightforward using a question in an exam paper to find out if students can recall the word equation for photosynthesis but not so easy using the same tools to see if they can prepare a slide for viewing under a microscope. Obviously, the nature of summative assessment can be changed; different assessment activities can be

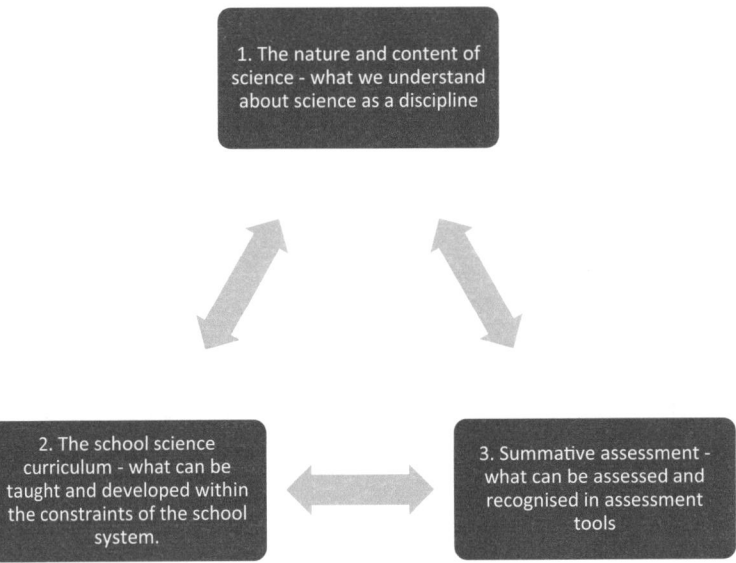

■ **Figure 17.1** Model showing relationship among the nature of science, the school curriculum, and what can be summatively assessed

deployed, and various types of questions can be included and this will determine assessed outcomes. The summative assessment does not necessarily have to be an unseen terminal examination paper; it could include the use of pre-release materials, project work, or the direct use of practical assessment. Coursework can take various forms and may contribute to the set of marks determining the results. However, the design of the course and the assessment tools specified will determine what is possible here.

In some cases the relationship between these elements is clear and straightforward. Many physicists would argue, for example, that Newton's Laws of Motion are a key concept in their discipline. Many educationalists would want to include an understanding of these in the school curriculum and their application can be assessed using questions in an exam paper. In other cases it may be more problematic. For example, many scientists work as part of teams alongside partners with expertise in other disciplines, both from other areas of science or from other subject areas. A key skill for scientists is therefore being able to contribute insights from their perspective in a broader project. This is fine to be recognised as part of working as a scientist but may not be so easy to reflect in a school curriculum, structured with discrete subjects. Some schools may find ways of organising cross curricular activities, but these are unlikely to influence exam grades. It would certainly be a challenge to write a question for an exam paper which assessed whether a pupil could work in such a way. Assessing whether the candidate was familiar with examples of such activity is easier but determining whether they possess the skill to perform such a role would be more difficult.

Regulatory authorities will want to reflect the nature of the subject in the assessment (i.e. to make the assessment and the outcomes generated as valid as possible) but there may be pressure to restrict the summative assessment to components that will lead to a high degree of reliability. It is easier to standardise and moderate marking of responses to an exam question than to manage coursework for example; it is also easier to write an exam paper to produce a spread of marks than to set up practical assessments to do the same.

Assessment objectives

Both teachers and pupils may assume that all the questions in an exam will have the function of ascertaining knowledge and understanding; this is often not the case. There is often a clear statement of intent on the part of assessment designers; this is typically done by the stating of assessment objectives. These are crucial as they determine the kind of questions that are asked.

For example in the GCSE Science specifications used for examinations from 2018 in England, Assessment Objective 1 (40 per cent of the marks) covered the demonstration by candidates of their knowledge and understanding of scientific ideas, scientific techniques, and procedures; Assessment Objective 2 (also 40 per cent of the marks) addressed the application of knowledge and understanding of scientific ideas, scientific enquiry, techniques, and procedures; and Assessment Objective 3 (worth 20 per cent of the marks) covered their proficiency in analysing information and ideas to interpret and evaluate; make judgments and draw conclusions; and develop and improve experimental procedures (Ofqual, 2015b).

When questions are being written, the examiner has to attribute every item (the components of a questions) to an assessment objective. The proportions in the previous example show that there will be as many marks going on questions that require students to apply their knowledge and understanding as are attached to the demonstration of that knowledge and understanding.

For example, if we ask pupils to explain the process of transpiration, that can be classed as a knowledge and understanding (AO1 using the taxonomy from earlier) question, assuming that this idea was clearly identified in the specification. If we want to ask students to apply their knowledge to another context we might, for example, ask them to look at a picture of a stingray and suggest how it is adapted to its habitat. This is an application (AO2) question because although the concept of adaptation might well be in the specification, that example isn't. We might also show pupils a table of data showing the gestation period and mean body mass of various female mammals and ask them to suggest what kind of pattern they can identify; this is an interpretation and evaluation (AO3) question. Good teaching involves asking different types of questions. This isn't just because examiners will be looking for proficiency in these areas but because this gets pupils thinking in different ways and developing their capacity as scientists.

The design of questions for summative assessment has clear implications for lesson design. If pupils are taught as if it is the demonstration of knowledge and understanding that is the prime determinant of success on the course (and, indeed, to being a scientist) then they may be unprepared either for the exams or for advancement within the subject.

Summative assessment often makes extensive and technical use of command words. Pupils are expected to be able to respond to these and to understand the different between, for example, describing and explaining. This may require both explicit teaching (it's not easy to understand what's expected in response to an 'explain' question for example if the notion of a causal link isn't clear) and frequent revisiting. The range of command words is greater than that which some pupils have experienced.

Care needs to be exercised here if teachers are designing summative assessments for a course from their own principles rather than based on those of awarding organisations. It can be argued that pupils (and, indeed, teachers) will draw conclusions from those about what is valued in the subject. For example an assessment that consisted entirely of a battery of multiple choice questions based on knowledge and understanding might lead pupils to think that science was very much about what is known and understood but less about application, interpretation, or indeed being able to develop a fuller explanation.

Task 17.1 Using assessment objectives to develop different types of questions

- Look at an example of a specification or syllabus that is used in schools that you have worked in and locate the assessment objectives
- Now look at some examples of questions that are used and see if you can attribute them to these objectives. Try to find at least one example of a question that fits each of the assessment objectives
- Now think about ways of developing students' ability to answer that type of question.

ASSESSMENT OF PRACTICAL SKILLS

Practical work has an important role in science education and its function is explored elsewhere in this book. It may also feature in summative assessment and it is important to understand how. This is a narrower definition than the broader area of processes that may be referred to as 'Scientific Enquiry'; by practical work we are talking about the set of skills needed to use apparatus to gather first hand evidence.

Assessment objectives also apply to practical skills. Pupils can be expected to know and understand certain techniques and they may also be expected to see how these could be applied to a range of contexts. Furthermore, pupils might be asked to critique a practical procedure and suggest how it could be improved.

There are, in essence, two ways of assessing such skills. One is to use direct assessment, in which the candidate is observed carrying out some procedure and is assessed. They might, for example, be asked to make a spring from a piece of wire, test the spring by loading it, and take readings of the length and then use the data to work out the spring constant. The assessor would note proficiency and make decisions about the performance demonstrated. It could be that the pupil is observed as they do this or they could make records that are then scrutinised, depending on the skill. The other way is to use indirect assessment, which usually means that although the pupil may still carry out the same activity, they will be asked questions in an exam at some later date relating to such work.

Each has merits. Direct assessment is likely to be the best way of seeing whether a pupil can physically make a spring, support it, and get it to carry weights. If assessing practical skills indirectly, an examiner might present a set of data and ask the candidate to identify a trend, spot outliers, and suggest a conclusion. Although this might feel a little artificial compared with getting pupils to look at their own data, it has advantages. The examiner can make sure that, for example, there are some outliers to comment upon (and also that there aren't too many); they can include graphs that are more or less challenging and which draw upon a much wider range of contexts than pupils could themselves produce from their own practical work. Note though that these skill sets are not identical, the manipulation of data is not the same as the manipulation of equipment.

Practical work is also one of the aspects of the science curriculum in which it is more likely that pupils will find themselves using a range of teamwork skills in order to succeed; this is a good example of an aspect of learning that is valued by employers but may not always be recognised in summative assessment.

The assessment of practical work has been the area of much debate and there is significant variation between practice in various jurisdictions. It was noted in the executive summary of a report on 'Improving the assessment of practical work in science' (Reiss, Abrahams and Sharpe, 2012, p. 4) that:

> Awarding bodies and others should consider carefully the optimum balance between the direct and the indirect assessment of practical work in science.
>
> Given that employers value skills such as team working, it may sometimes be appropriate, as in the assessment of drama, to use practical work in science to assess students' collaborative as well as individual skills.
>
> Greater use of teachers should be made in the summative assessment of their students' practical work, accompanied by a robust moderation procedure.

In the report 'Good Practical Science' (Holman, 2017) there are five purposes suggested for practical activities:

A: To teach the principles of scientific enquiry

B: To improve understanding of theory through practical experience

C: To teach specific practical skills, such as measurement and observation, that may be useful in future study or employment

D: To motivate and engage students

E: To develop higher level skills and attributes such as communication, teamwork and perseverance.

Holman, 2017, p. 17

The first two of these suggest that a teacher can usefully use practical work to promote outcomes likely to be recognised in exams. The third may have such an impact, depending upon the role of and assessment of practical skills. GSE science courses in England have, for example, a practical endorsement component which recognises and certifies these skills. The fourth point has a general impact upon student performance but the fifth refers to an aspect which, although it may be acknowledged in exam specifications, is often not recognised in assessment activities or used to influence final grades. This will be picked up again later in this chapter.

Task 17.2 **Considering how to assess practical competencies**

■ Take a practical activity such as, for example, extracting salt crystals from sea water and determining the percentage yield. Consider the various skills that are involved with this; these might include ones relating to manipulation of equipment but also ones linked to performing calculations and evaluating the procedure. Include skills such as collaboration and problem solving in your reflection.

■ Consider how these could be effectively assessed and how these would lend themselves to direct or indirect assessment.

■ Research how a specification or syllabus used in a school you have worked in assesses practical skills.

ASKING MORE CHALLENGING QUESTIONS

What makes hard questions hard? What should have become clear by this point is that summative assessment is more than just a battery of questions on different topics. Summative assessments are designed to assess not just coverage but also various levels of competence, so questions will often be pitched at different levels of challenge.

What determines the level of demand of a question is a complex set of factors. There are a number of ways of varying the degree of challenge in questions. It is commonly understood that some contexts are more straight-forward than others. For example, it is probably easier to understand wave reflection than wave refraction. The idea that waves bounce off a reflective surface is supported by concrete experiences from everyday life with various objects and the relationship between the angles of incidence and reflection is a straightforward one. Refraction is harder to grasp. An examiner looking to write a harder question might use refraction as a context and the question would be more difficult. The level of demand of the science concept has been increased.

However, it isn't the only way in which questions can be made more dif-ficult. The examiner might stay with the topic of reflection but to ask the candidate to apply it to a different context. They might, for example, include a diagram showing a bicycle reflector and ask for an explanation as to how this works, possibly asking the student to compare it with a mirror and to jus-tify why one is better than another as a safety device. It's still reflection, but it's less straightforward. The context of the application has been made more challenging.

Alternatively, the examiner might vary the extent of cueing. A question might signpost clearly the context that is being used or it may be that the candidate has to do more work to understand and interpret the context, if fewer cues are offered.

Many exam questions consist of a series of items and often these items will be on a common theme. These items may be sequenced because the question writer is using an idea called 'ramping' (see also chapter 16). The items start off being more straightforward and then, as the student progresses through the question, the later items become harder. For example, in a question on photosynthesis, the first item might ask the candidate to recall the word equa-tion for photosynthesis, the second one might be based on why the quanti-ties of products vary over a 24-hour cycle, and the third one might ask how an experiment investigating the process could be set up. This could then be followed by another item which included a set of sample data and asked the pupil to interpret this and suggest a conclusion. Again, this is good practice in general teaching, using a range of questions and managing them to move pupils from understanding an idea to applying it and being able to interpret new evidence.

Task 17.3 **Developing challenging questions**

■ Take a topic you are familiar with and write a question that would elicit some understanding of some part of it. Don't use one that is closed and could be an-swered in a single word but rather one that will require some thought. It should, however, be fairly straightforward

■ Now try rewriting this question using the idea of ramping to come up with three other versions, each more challenging but in different ways.

PREPARING STUDENTS FOR SUMMATIVE ASSESSMENT

It is appropriate here to consider the implications of taking a group of pupils through a course over several years and preparing them for examination at the end of it. It is unlikely to be effective to assume that your job is to teach them the material in the first place but that revisiting this and keeping their understanding fresh is their responsibility, needing no support or guidance from you. There are practical strategies that can be used effectively to support pupils to retain knowledge.

When something has been learned it will be held in the working memory but needs to be transferred to the long-term memory; this is referred to as encoding. Before it can be accessed and applied, it needs to be retrieved from the long term memory and brought back into the working memory. An important function for teachers is to support the development of this encoding and retrieval. The use of models from cognitive psychology in teaching is explored in more detail in Chapter 9.

One of the useful ideas in this area is the notion of spaced repetition (see Figure 17.2). One of the ways of addressing this is to revisit it after a period of time. A pioneer in this area was Paul Pimsleur, who researched ways of supporting adults to learn vocabulary when becoming proficient in another language. He found that the most effective way of doing this was to revisit the material but to leave increasing gaps between refreshing the learning. In other words, you might come back to something a week later, a month later, and a term later. Sometimes, of course, this can be done as an integral part of the course; it may be appropriate to revisit ideas about energy transfer and efficiency when exploring the operation of the transformer. Sometimes the revisit doesn't have an obvious role in the topic being studied but it's important to do it anyway. Teaching pupils something when they're 14 and expecting them to remember it when they're 16 is unrealistic. These revisits need to be planned if they're going to happen.

A more sophisticated system was developed by Sebastian Leitner. He proposed that when the revisiting takes place, the learner sorts the items into those successfully recalled and those not (see Figure 17.3). The ones less successfully learned will be revised again sooner whereas those more successfully learned will be pushed back in the plan.

Another important idea is that of retrieval practice. This has a strong relationship with spaced repetition but focuses particularly on how ideas are refreshed.

■ **Figure 17.2** The intervals recommended by Pimsleur. Within the context of a typical secondary school timetable, the first four or five could be within the lesson and the last four within subsequent lessons. However, the general principle is increasing intervals in order to promote longer term memory

Source: 'Spaced repetition', https://www.immagic.com/eLibrary/ARCHIVES/GENERAL/WIKIPEDI/W110427S.pdf

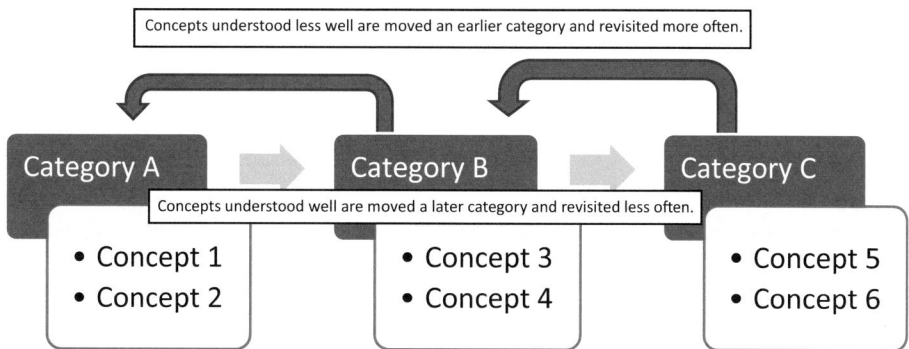

| Concepts understood less well are moved an earlier category and revisited more often. |

Category A → **Category B** → **Category C**

Concepts understood well are moved a later category and revisited less often.

Category A	Category B	Category C
• Concept 1 • Concept 2	• Concept 3 • Concept 4	• Concept 5 • Concept 6

■ **Figure 17.3** A diagrammatical representation of one way of implementing Leitner's approach. Concepts 1 and 2 are ones that students don't find it easy to recall and use so they are in Category A and are revisited more frequently. Concepts 3 and 4 have been mastered more successfully so come up less frequently; 5 and 6 are least problematical and so come up least often. Concepts are moved from one category to another according to evidence of performance. Different versions of the system have different numbers of categories. A common approach, especially if this is being used by individual students, is to have the concepts on individual flash cards and move them between boxes, which represent the categories

Source: The Leitner Flash Card System: https://virtualsalt.com/learn10.html

Good practice includes using a range of strategies. An obvious one is the use of practice exam questions; these will often involve pupils in recalling material from earlier lessons. There will be other opportunities too, and good teachers will look out for these and use them in questioning strategies. For example, learning about the electric motor is a good opportunity to ask pupils to recall ideas about current flow, forces, and magnetic fields. Similarly, understanding an ecosystem provides a context for questions about photosynthesis, transpiration, and food webs. However, what retrieval practice also highlights is the importance of pupils retrieving ideas in different ways. If pupils work in different ways to recall and apply material previously learned, it will strengthen long term memory and the capacity for retrieval.

Task 17.4 **Putting ideas about spaced repetition into practice**

■ Look at a particular topic that you are teaching or planning to teach. This should be one in which understanding will be assessed in summative assessment at a later date. Try to work with an example that would be featured earlier in a course
■ Consider the long term plans for this course and identify opportunities within later topics to revisit your chosen set of ideas. In some cases this will be fairly straightforward but in other cases there may be no obvious opportunity and so a specific revisit session will be needed
■ Now think about ways of revisiting that idea. This shouldn't be the teacher re-teaching it but pupils being challenged and supported in different ways to recall it.

BROADER AIMS OF SCIENCE EDUCATION

Summative assessment will tend to foreground certain aspects of the subject and place less value on others. We may want to retain at least some of those in the learning experiences that our pupils experience because they are valuable and authentic. Depending on the jurisdiction in which you are working, the nature of summative assessment may vary and the aspects of science education which are reflected in the associated summative assessment may be different. However, you may find that there are a number of skills which are not reflected in summative assessment. For example, STEM Learning have identified three useful groups of employability skills. The first of these focuses upon how the pupil works, looking at aspects such as their organisational skills, ability to work under pressure, being able to learn, and to use their initiative. The second focuses upon how they work with others, including teamwork, negotiation, and communication but also their capacity to value diversity. The third area considers how they think and includes aspects such as numeracy and problem solving.

Some forms of summative assessment attempt to assess at least some aspects of problem solving and many aspects of numeracy, but it's harder to see how some of the others could be included if a course is assessed entirely by formal written papers. Few teachers would argue either that the full set of skills weren't important or relevant or that they didn't want to try to develop those (given time and resources). Furthermore, it is not unusual to hear of commercial STEM organisations underlining the importance of such skills.

We want our pupils not only to get good grades but also to secure places in subsequent education, training, and employment. It is therefore in our interests to explore ways in which such skills can be developed and, at least informally, recognised. However, we also want all pupils to be scientifically literate and to have a good sense of how to evaluate claims and use evidence. This is important and is worth emphasising. Teachers are sometimes challenged by students as to why they should be studying science; responses along the lines of 'science is all around us' are less persuasive than we might imagine them to be. Pupils know this but they may not see science as being for them. Justifying it on the grounds of potential employment is only part of the answer. People need to be able to use scientific reasoning to consider issues in their lives such as the case for vaccination and their role in responding to climate change. Enabling young people to do this effectively is a good justification for science being a core subject. Much valuable research into students' attitudes towards science has been carried out as part of the ASPIRES project (see e.g. Archer et al, 2015a and Archer et al, 2015b).

A good opportunity to develop at least some of these is when pupils are involved in practical work. This depends upon the nature of the activity, the extent to which there is scope for decision making, and the role of collaboration, but for the teacher looking to develop this dimension there is scope. However, this is not to say that these will be addressed simply by doing practical work. It needs conscious planning and focused delivery for them to be a strong part of teaching and learning.

Task 17.5 **Developing key employability skills**

■ Select one of the employability skills discussed in this chapter
■ Consider how you could incorporate the teaching of this skill into a science lesson. What kind of activity might lend itself well?
■ How would you know if your pupils were successfully developing this skill?

USING SUMMATIVE ASSESSMENT DATA

Performance data can be used to evaluate both pupil performance and the effectiveness of the curriculum. This information could be gathered from a mock exam or end-of-year assessment which has the broad characteristics of the summative assessments.

One obvious way of doing this is to analyse responses at question level, either by individual student or by whole group. The class might have done better on Q1 of a mock exam paper than on Q2 and it would be appropriate to look at the topics they were based on. The Q2 topic would probably be a better bet to return to as the evidence might suggest their grasp of the underlying ideas was weaker. However, caution needs to be exercised here as there are many reasons why students might underperform, and not understanding the concept is only one of them. It might be that Q2 required pupils to apply an idea to an unfamiliar context and this was something that pupils weren't used to doing or it might be that it required the interpretation of a graph and that this was a skill in which candidates were weaker. Of course, producing a high quality response requires quite a collection of skills and teachers only have so much time for analysis, so there will be a limit as to what can be done. However, it is worthwhile exploring how this can be taken a little further.

One way is to look at particular aspects such as assessment components, mathematical skills, or practical skills. A question based on a particular experiment might have caused problems because pupils hadn't understood that specific activity but it is worthwhile to see if it was a more generic issue. Did pupils always underperform on items related to practical work? Did this need to be a focus for revision or even for reviewing the way that practical work is taught? It might be that pupils satisfactorily carry out the experiments but are not good at translating that into high quality responses on exam papers.

It is relatively straightforward to analyse pupil performance against assessment objectives. Pupils who score well on knowledge and understanding questions but do less well on application or on interpretation and evaluation questions, for example, need the teaching and the intervention to address this. This is not an argument against topic-based question level analysis. Sometimes pupils underperform because they don't understand a particular concept and it needs to be revisited. However, this isn't always the case and being open to other reasons is important.

Task 17.6 **Analysing summative assessment data to inform teaching**

■ Locate an example of an assessment activity that has been used in the science department you are working in and which consists of a number of exam style questions. It might, for example, be an end-of-term test or a mock exam paper

■ Think about various ways in which pupil performance on this might be usefully analysed to direct future teaching and learning. You might consider such factors as:

■ The topics featured in the questions which pupils found more challenging

■ Their performance against assessment objectives. This would compare performance in, say, knowledge and understanding questions with those assessing application

■ Particular skills such as calculation or interpretation of graphs

■ Now talk to a member of the teaching team who used this assessment and ask them to describe what analysis they did and how useful it was.

SUMMARY AND KEY POINTS

Summative assessment has a strong influence upon teaching and learning in science because performance in exams is seen to be important. This influence might be progressive; depending upon the design of the specification and exams it may encourage teachers and pupils to move away from a purely knowledge-based approach to the subject and to see the importance of skills such as application, interpretation, and evaluation. It may also encourage the development of skills such as retrieval practices and being able to develop knowledge and understanding over a period of time.

In other ways exams are a less benign influence. There are a number of skills which, though they may be seen by higher education and employers to be valuable, may not be reflected in the assessment tools and therefore run the risk of being underrepresented or undervalued in the curriculum. It is appropriate as professionals to consider these and decide whether and how they may be incorporated into our classroom practice.

It is important, however, for teachers to understand the design of examinations, what they are designed to recognise, and how they go about doing that. By so doing, teachers will be more able to prepare pupils to perform as well as they can. Furthermore, they will also be able to analyse data to diagnose performance and to plan teaching and intervention effectively.

FURTHER RESOURCES

Awarding organisations may have resources available to support a better understanding of the nature and purpose of summative assessments on offer. It is worthwhile to look at the structure of the specification and become familiar with what's in the appendices as well as the main content sections. Relevant sites for the UK include:

England:

- AQA: www.aqa.org.uk/subjects/science
- Cambridge International: www.cambridgeinternational.org/programmes-and-qualifications/cambridge-igcse-science-combined-0653/
- Edexcel: https://qualifications.pearson.com/en/qualifications/edexcel-gcses/sciences-2016.html
- OCR: www.ocr.org.uk/subjects/science/
- Eduqas: www.eduqas.co.uk/qualifications/combined-science-gcse/#tab_overview

Northern Ireland:

- CCEA: https://ccea.org.uk/science

Scotland:

- SQA Sciences: www.sqa.org.uk/sqa/45718.html

Wales:

- WJEC: www.wjec.co.uk/qualifications/science-double-gcse-award#tab_overview

Holman, J. (2017). *Good Practical Science*. The Gatsby Charitable Foundation. Online, available at https://www.gatsby.org.uk/education/programmes/support-for-practical-science-in-schools

This is a useful and well developed summary of the factors promoting effective practice. It is worthwhile to consider the benchmarks and their relationship with the components of summative assessment being used in your school. Science departments sometimes have to work hard to justify the allocation of resources and it is useful to reflect on aspects that can be justified with reference to likely immediate gains in student outcomes in summative assessment and those linked to broader aims in science education.

CogSciSci. https://cogscisci.wordpress.com/

Run by teachers and other educational professionals, the aim of this helpful site is to share effective cognitive psychology practices that relate to the teaching of science.

The Learning Scientists. www.learningscientists.org

Learning scientists is a group of practitioners interested in cognitive science and how key findings can support effective learning; their site includes a number of strategies ready to explore and use.

INCLUSIVE AND ADAPTIVE SCIENCE TEACHING

Darren Moore

INTRODUCTION

Think about the last science lesson that you taught or observed:

Was it inclusive?
Was the teaching adapted to meet the various needs of pupils?

Your answer may well be 'probably' or 'yes', but please consider now how that lesson could have been more inclusive, or if there were other relevant pupil needs that the teaching could have further responded to.

This chapter will help you formulate some answers to these taxing questions. Inclusion – as it pertains to education – and adaptive teaching are important terms but notoriously difficult to pinpoint (Schipper et al., 2020). Science can be a challenging subject to learn, as well as teach. It is therefore important that your science teaching is accessible and engaging, as well as responding to the needs of individual pupils. However, it also is important that this is done in a way that does not make the work of planning and preparing your lessons too oner-ous. Fortunately, much of the advice in this chapter and in the wider literature is inclusive itself. It is often the case that a resource that caters for the needs of one pupil can also benefit the learning of other pupils, and an activity that might make science more accessible for one pupil also does the same job for their peers. Moreover, the idea of 'quality first teaching' as the starting point for meeting a wide range of pupils' need is recommended in research, policy, and practice (Lowe & Joffe, 2017).

After considering how we might define some important terms like inclusive teaching, differentiation, and adaptive teaching, the remainder of the chapter gives some practical examples for your science teaching. This chapter draws upon research evidence about special educational needs and disabilities (SEND)

DOI: 10.4324/9781003110187-19

and science teaching and learning. That is an area in which I am interested and have supported student teachers myself. But the focus on pupil needs could have easily drawn upon literature on teaching pupils for whom English is not their first language, ethnic minority pupils, vulnerable pupils, or disadvantaged pupils. It is therefore worth keeping in mind some of these other pupil groups as you think about issues of inclusive science teaching, particularly as we finish the chapter with an examination of some wider issues for science teaching.

OBJECTIVES

At the end of this chapter you will be able to:

- Appreciate the range of definitions of inclusion
- Define adaptive teaching and recognise its links to differentiation
- Consider practical ways of adapting your teaching for pupils with special educational needs and disabilities
- Describe some key issues for inclusive science teaching

DEFINING INCLUSIVE TEACHING

Approaches that treat all learners the same might appear to be more inclusive as everyone is treated equally. However, this may overlook individual differences and learning needs. Individual approaches may meet individual needs; however, these approaches may come at the expense of the separation of learners and thus emphasising differences. Norwich (2010) refers to this as the dilemma of difference. Inclusion as a general guiding principle should strengthen equal access to quality learning opportunities for all learners. Rouse (2008) suggests that

> developing effective inclusive practice is about not only about extending teachers' knowledge, but it is also about encouraging them to do things differently and getting them to reconsider their attitudes and beliefs. In other words, it should be about 'knowing', 'doing', and 'believing'.
>
> (Rouse, 2008, p. 12)

Rouse builds on this, using the elements of head, hand, and heart, which I have elaborated in terms of the messages for student teachers in Figure 18.1.

Inclusive pedagogy is often defined in wide-ranging terms like this and often associated with special educational needs. But the messages are relevant for teaching all learners. It refers to the development of learning opportunities that are made available for *all* pupils, rather than *most* with different options for a few learner needs. Inclusive pedagogy means that all pupils can participate in lessons (Florian & Black-Hawkins, 2011).

Head – Teachers' knowledge and skills. One way to improve your inclusive practice is to be prepared to know when, why and how to respond when children experience difficulties in learning.

Hand – Teachers' actions. How will you do things differently, try new things? For example, to try working collaboratively with others - pupils; teaching assistants; parents; specialists etc.

Heart – Teachers' attitudes, values and beliefs. View and deal with difference as part of the human condition, rather than as a difference/a problem to be ameliorated. Share this view with pupils.

■ **Figure 18.1** Rouse's head, hand, and heart view of inclusive practice applied to student teachers

---

Task 18.1 **Head, hand, and heart targets**

Set yourself an improvement target under each of Head, Hand, Heart. Formulate a plan to address the targets. Share the targets and plan with your mentor. For instance, for 'Head' you may aim to anticipate difficulties in the content in your next unit (Chapters 9 and 11 will provide ideas for addressing this).

ADAPTIVE TEACHING AND DIFFERENTIATION

Differentiation is perhaps easier to define than inclusive pedagogy, with Armstrong et al.'s (2010) definition explaining that 'differentiation is concerned with delivering education in different sorts of ways for 'different' sorts of student' (Armstrong, 2010, p. 99). However, things quickly become complicated when considering how differentiation is often broken down into types: for instance, differentiation by content, process, or outcome. In England, for example student and early career teachers must show they can:

> Adapt teaching to respond to the strengths and needs of all pupils . . . know when and how to differentiate appropriately, using approaches which enable pupils to be taught effectively.

> (DfE, 2011, p. 11)

But then the Ofsted Education Inspection Framework (Ofsted, 2019) says:

> They [teachers] respond and adapt their teaching as necessary, without unnecessarily elaborate or differentiated approaches.

> (p. 9)

Webster and Blatchford (2019) note how broadly differentiation can be interpreted by a range of school secondary school staff, but at the same time how narrow and under-conceptualised some of their definitions are. Fortunately, researchers have tackled the challenges of both conceptualising differentiation and considering the impact of differentiated teaching. Van Geel et al. (2019) recognise that differentiation is not only about the decisions made to adapt teaching and learning for certain individuals or groups, but rather there are a range of teaching skills that span across the cycle of plan, do, reflect. A medium-term plan or scheme of learning will not specify differentiation, but drawing on both this broad plan and recent individual assessment, teachers are ready to plan the teaching approach and goals and consider pupil needs. Preparing the actual lesson involves determining any grouping and selecting material (both provide opportunities for adapting to pupil needs). In providing adapted instruction it is important to monitor progress and fit the teaching to meet pupil needs. Reflection on the lesson will note individual and group needs for forthcoming lessons, as well as importantly consider the value of specific adaptive teaching techniques – what was worth the effort?

Rigorous reviews find that relatively little research has assessed the impact of differentiated instruction in secondary-level education (Graham et al., 2021; Smale-Jacobse et al., 2019). The research tends to show some benefits for a wide range of differentiation including attainment grouping, individualising instruction, mastery learning, and additional support. These empirical findings give some indication of the possible benefits of differentiated instruction. However, they also point out that there are still severe knowledge gaps. This again indicates the wide range of teaching strategies that may be employed in relation to differentiation or adaptive teaching and that we do not know whether a particular strategy is more effective than others. The reviews of research indicate that across differentiated instruction, appropriate level of challenge for pupils is important. There is also a small amount of evidence that differentiated teaching is more beneficial for lower attainers (Smale-Jacobse et al., 2019). Differentiation involves a holistic approach to help achieve more inclusive teaching, rather than being about different pupils getting different offerings (Lindner & Schwab, 2020), with the important caveat that adaptive teaching is not far removed from the high quality teaching for all pupils previously mentioned.

Task 18.2 **Approaches to differentiation for your science teaching**

Read the 'Literature Synthesis' section of Smale-Jacobse et al. (2019) https://www.frontiersin.org/articles/10.3389/fpsyg.2019.02366 (full text of this article is open access). Consider which of the approaches to differentiation are more or less relevant for science teaching, what barriers there are to implementing relevant approaches in your science lessons, and which approaches you are already using. Compare your notes with another student teacher.

TEACHING SCIENCE TO LEARNERS WITH SPECIAL EDUCATIONAL NEEDS AND DISABILITY

It is has been said that 'there is very little in the way of dedicated subject-specific continuing professional development (CPD) provision for science teachers in special schools, nor for those who work in the mainstream but have children with special educational needs in their classes' (Bullough & Booth, 2013, p. 12). This section of the chapter aims to challenge the idea that there will be one best way to teach science to a pupil with dyslexia or autism and instead help you consider how resources and ideas for pupils with SEND from research literature may benefit a wide range of learners in your classes get to grips with science.

Definitions of SEND vary widely across the world. Some countries define SEND using a general definition that might refer to disabled pupils or those with significant learning difficulties. Other countries, such as Japan, define it by categorising pupils with SEND into more than ten different categories. A simplified definition is:

> A pupil who is not able to make progress in the education made generally available for all pupils of their age, without significant additional support or adaptation.

What is in scope under this definition does vary across countries. For instance, Turkey includes gifted and talented amongst SEND, Switzerland includes a foreign first language and, at the time of writing, England organises SEND provision under four categories (Department for Education (DfE) and Department of Health (DoH), 2015), which are:

■ Communication and interaction such as: autism spectrum disorder (ASD); speech language and communication needs
■ Cognition and learning such as: dyslexia; moderate learning difficulty
■ Social, emotional, and mental health difficulties such as: attention-deficit/ hyperactivity disorder (ADHD); anxiety
■ Sensory and/or physical needs such as: visual impairment; cerebral palsy.

Different categories of SEND can lead to the belief that there are teaching strategies to be learnt specific to supporting learners with ADHD or cerebral palsy. While it is important to understand frequently occurring SEND (see the condition specific videos in the further resources section), there are three reasons why searching for strategies specific to each condition is unhelpful. First, there are wide individual differences among pupils that have a particular conditioning. Functioning will vary amongst pupils with autism, with mainstream settings being inappropriate for some, while research has suggested that most people with autism have other co-occurring conditions. Second, there is also overlap across conditions. To take an example with a relatively rare prevalence like Foetal Alcohol Syndrome, because difficulties for young people can affect focus, memory, communication, and motor skills amongst others, comprehensive teaching advice that purports to be for this condition contains strategies that are very relevant for a range of other SEND too (e.g. NHS Ayrshire & Arran, 2019). Third, the SEND Code of Practice for England referenced earlier recommends high quality personalised teaching as the starting point to meet the individual needs of many pupils. Moreover, Davis et al. (2004) conducted a literature review and concluded that there was no single strategy for teaching pupils with SEND, and there were not recommended methods for particular groups of pupils with SEND.

In 2020 the Education Endowment Foundation (EEF) published their guidance report on supporting SEND in mainstream schools. It is notable how the recommendations fit the previous discussion. If you view the summary of recommendations poster here https://educationendowmentfoundation.org.uk/education-evidence/guidance-reports/send you will see that only the fourth recommendation, to complement high quality teaching with small group and individual interventions, is recommending action that moves beyond responding to the needs of all pupils in science.

Task 18.3 **EEF SEND guidance report recommendations**

Choose one of the five recommendations from the EEF SEND Guidance Report (Education Endowment Foundation, 2020) about which you would like to find out more practical teaching advice. Then read that section of the Guidance Report https://educationendowmentfoundation.org.uk/education-evidence/guidance-reports/send

While reading about your recommendation make a note of:

1 a strategy or principle that you would seek to implement in your future practice
2 something that would be challenging for you as a teacher to implement
3 a reference to follow up later

SEND TEACHING STRATEGIES IN SCIENCE

In this section, we will consider more specific research about working with pupils with SEND in science lessons. I invite you to consider how you might be able to use these ideas in your science teaching and not necessarily with a focus only on pupils with SEND but potentially for a whole class.

Science writing heuristic approach

Villanueva and Hand (2011) focused on *scientific literacy* as an intrinsic goal of science education to overcome potential challenges when working with pupils with a range of SEND, in particular processing and cognition difficulties exhibited by learners with dyslexia. Science writing heuristic (SWH) is a well-developed writing-to-learn model for learning from practical activities in secondary science and can be used by teachers as a framework from which to design classroom activities. It is designed to encourage construction of conceptual knowledge. It is also based on relationships among questions, evidence, and claims. SWH was developed by Carolyn Keys and colleagues (Keys et al., 1999). Pupils complete writing according to a template, structured to consider the processes involved in the activity, with particular emphasis placed on claims, evidence, and reflection. There is evidence that SWH helps students to generate meaning from data and to make connections among procedures, data, evidence, and conclusions. It also promotes metacognition and serves as an example of the difference between more metacognitive approaches to practical science versus perhaps a more typical process of observing a demonstration, completing the same practical and then comparing results.

SWH is intended as a teaching model of two parts, one for the teacher actions and the other for pupil activities. A colleague and I have adapted it as follows. The left-hand column represent ways of structuring teaching and the right hand column provides an individual writing frame for pupils to complete as they work through these activities:

■ **Table 18.1** A template to promote scientific literacy (adapted from Moore & Black, 2021, p. 297)

	Teacher designed activities to promote scientific literacy	Writing frame prompts for pupils
1	Exploration of existing understanding through individual or group concept mapping	Beginning ideas: What questions do I have about this topic?
2	Pre-practical activities, including informal writing, brainstorming, posing questions, and planning	Tests: What will I do and why?
3	Pupils' participation in practical activity	Observations: What did I see?
4	Negotiation phase 1: writing personal meanings for practical activity (noting observations)	Claims: What can I claim?
5	Negotiation phase 2: sharing and comparing data interpretations in small groups (for example making group charts)	Evidence: How do I know? Why am I making these claims?
6	Negotiation phase 3: comparing science ideas to textbooks or other printed resources (how do findings fit hypotheses?)	Reading: How do my ideas compare with other ideas?
7	Negotiation phase 4: individual reflection and writing (for example creating a presentation such as a poster or report for a larger audience)	Communication: How can I communicate my claims, evidence, readings, findings, and evaluations with my peers/class/teacher?
8	Exploration of post-teaching understanding through concept mapping	Reflection: How have my ideas changed?

Task 18.4 **Applying the science writing heuristic approach**

Take a practical lesson that you have delivered or observed in practice and make notes on how you could use the writing frame to structure pupils' learning and thinking during the practical. **Reflect** on whether you would use something like this with all pupils and any challenges that might be faced applying it to practical lessons. Refer back to Chapter 13, Practical Work, to consider how this might build upon your planning for the actual delivery of practical work in science.

Graphic organisers

Graphic organisers are used to organise knowledge, concepts, and ideas. Examples include Venn diagrams, T-Charts of pros and cons, mind-maps, cognitive maps, semantic maps, and chronologies or event chains. They can be effective tools for supporting learning. Dexter et al. (2011) reviewed evidence for their use with learners with learning disabilities and found that use of graphic organisers was associated with increases in vocabulary knowledge, comprehension, and inferential knowledge. An example from the EEF guidance report on SEND teaching in mainstream schools is given on page 23, where a teacher might notice that a pupil is struggling to precisely define and understand what a 'planet' is. It uses a type of graphic organiser called the Frayer model as a flexible tool for this purpose. A Frayer model categorises information about a phenomenon according to: its definition, characteristics, examples, and non-examples. Figure 18.2 shows an example of a Frayer model for arteries. Pupils could be given a diagram like this as a scaffolding strategy or develop their own.

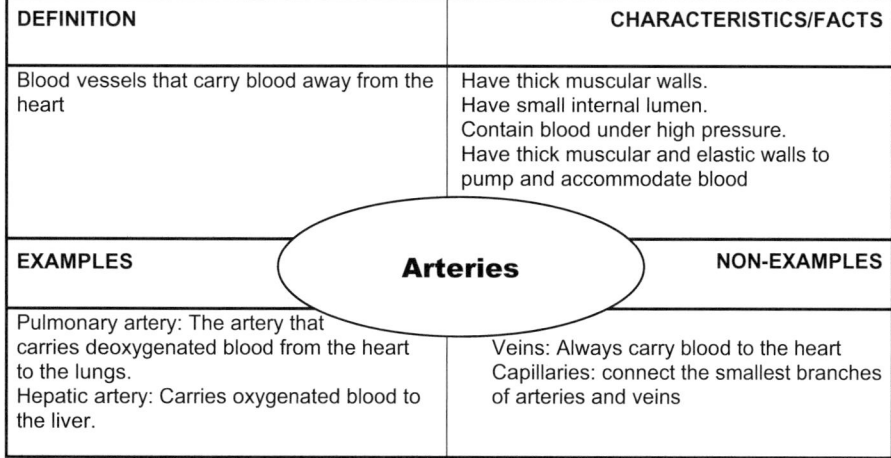

DEFINITION	CHARACTERISTICS/FACTS
Blood vessels that carry blood away from the heart	Have thick muscular walls. Have small internal lumen. Contain blood under high pressure. Have thick muscular and elastic walls to pump and accommodate blood
EXAMPLES	NON-EXAMPLES
Pulmonary artery: The artery that carries deoxygenated blood from the heart to the lungs. Hepatic artery: Carries oxygenated blood to the liver.	Veins: Always carry blood to the heart Capillaries: connect the smallest branches of arteries and veins

Arteries

■ **Figure 18.2** Frayer model for arteries

> ## Task 18.5 **Complete your own Frayer model**
>
> Take a lesson topic where you know that pupils may struggle with the concepts; it doesn't need to be pupils with identified SEND needs. Complete a brief Frayer model for that topic. **Reflect** on whether you would use something like this with all pupils, not just those with SEND and whether completing the model would be a useful activity for pupils to do themselves (perhaps with scaffolding for some learners).

Active learning in science

Rizzo and Taylor (2016) conducted a review of inquiry-based learning for pupils with SEND in science lessons and concluded that this type of teaching and learning can improve science achievement for pupils with SEND as much as for non-SEND pupils. This might be surprising, as there is a misconception that inquiry-based learning instruction is not appropriate for pupils with SEND, although Starling et al. (2015) found that a combination of inquiry-based practical activities and direct instructions improved performance among pupils with learning disabilities or ADHD, as compared to using either method alone. They suggested that any opportunity to teach science to pupils with SEND outside the classroom improves attitudes towards the learning of science by pupils with a wide range of learning needs but also enhances pupils' interest in science and their feeling of competency in the subject. Rather than visiting a science fair or taking a school trip (as useful an experience as they might be), consider opportunities for learning outside the classroom. Embedding active learning can be challenging, but it is important, as it provides an opportunity to connect all pupils with science without the challenges that a traditional classroom setting may pose pupils with SEND.

Szyjka and Mumba (2009) highlight that active learning can be a barrier for pupils with SEND when they are often given a very structured form of one-to-one instructions, which can limit these pupils' participation in active learning with (guided/open-ended) instructions and group work. As such active learning can be seen to promote more inclusive science teaching. Szyjka and Mumba (2009) suggest the use of scaffolding approaches for all pupils to facilitate more active learning approaches in science lessons.

Summary of SEND and science teaching

Good science teaching for pupils with SEND helps to include them in the learning experienced by the wider group. Teaching is adapted to support these learners but often the responses will be very relevant for a wider range of pupils who have a variety of needs and require support with their literacy and working memory. Here is a list of generally useful approaches for supporting pupils with SEND, as well as other pupils too:

Break activities into small, focused steps
Keep instructions short and concise
Simplify or shorten each class activity
Allow extra time or ensure that activities can be completed later
Use visual and concrete examples to aid understanding
Ask pupils to repeat instructions to clarify their understanding and explain to others
Provide alternative methods of recording responses, e.g. labelled pictures, flow-charts, writing frames, and diagrams
Ensure repetition and revisiting of key concepts over time
Praise progress, effort, and demonstration of scientific skills.

At times the nature of science teaching and learning means we will need to differentiate learning for particular pupil needs. We may need to support pupils with **sensory or physical difficulties** in accessing experimental work and specialist items. For example, pupils with hearing impairments may need support in using oscilloscopes and sound level meters; visually impaired pupils may need time and help to access activities using microscopes, in making observations (reading a thermometer, for example), or in interpreting graphs. We should be perceptive to the needs of pupils with **emotional and social difficulties** by providing structure and routine in practical situations. Pupils with **communication and interaction difficulties** may need additional help in accessing science-specific vocabulary and will appreciate the use of clear and concise language. Pupils with Autism Spectrum Disorder may be aided by structured routines in the collection, use, and tidying away of practical equipment. Pupils with moderate learning difficulties may need more contextual information for practical work or the work broken down into smaller chunks – looking at Key Stage 2 schemes of work can be helpful here.

KEY ISSUES FOR INCLUSIVE SCIENCE TEACHING

This chapter finishes with three areas that have the potential to challenge inclusive science teaching. First, the gender imbalance in the take up of further science study once science is no longer a compulsory subject. Secondly, the related issue of pupils' changing attitudes to science. Finally, I consider how teaching assistants can support teaching and learning in science lessons.

Gender gaps in science learning

The Institute of Physics (IoP) have completed a lot of important work focused on the gender imbalance of physics A-level and on issues around equality of education. Girls are under-represented among those who choose to study physics beyond the age of 16, in spite of there being no gender differences in results at GCSE level. The IoP is involved in a number of initiatives to tackle the expectations, stereotyping, and conscious and unconscious biases which combine to influence girls' and boys' choices of subjects. Their guide 'Opening Doors' presents good practice in relation to countering gender stereotyping in schools.

Task 18.6 **Reflecting on the Institute of Physics' inclusive learning checklist**

The IoP's work on gender balance has led to a toolkit and resources. One resource forms the basis of this activity. They have a checklist for teachers named 'Inclusive learning in the physics classroom – a checklist for teachers of physics'. This checklist provides some prompts for gender inclusive-practice. It can be located here: https://www.iop.org/sites/default/files/2019-06/Improving-gender-balance-inclusive-checklist.pdf

1 Complete the checklist in relation to your experience teaching across the sciences. Make a note of any areas for further development that you might target
2 Consider which of these prompts either relate to inclusive teaching more widely than gender as written or could be easily amended to consider inclusion more widely
3 Reflect on the extent to which these prompts provide a checklist for inclusive science teaching. What would need to be added to consider the breadth of pupils who need to be included in science, as well as the breadth of science?

Reflecting on this task should lead to a range of actions that you can put in place to help your science teaching be more inclusive.

Attitudes towards science

Research has reported that at the end of primary school, although the majority of children enjoy school science, have parents who encourage studying science, and hold positive views about scientists and science activities, very few aspire to a career in science (DeWitt et al., 2013). Children's attitudes towards school science start to decline once they are in secondary school. Although girls prefer science lessons at age 12, boys are more likely to aspire to science careers. Disadvantaged pupils and white pupils are also less likely to aspire to science careers. Other research has suggested that addressing these attitudes should be done by addressing interpersonal interaction between teacher and pupil, the relevance and authenticity of the topics being studied, and the diversity of the teaching methods (Raved & Zvi Assaraf, 2011). This suggests there is much that science teachers can do to shape attitudes and engage pupils.

Science capital is a concept that can help us understand patterns in science participation. It can help to explain why certain groups mentioned remain underrepresented in post-16 science and why many young people do not see science careers as for them (Godec et al., 2017). The concept of science capital can be thought of as like a belt for tools, containing what you know about science, how you think about science, who you know with science knowledge and experience, and how much science is part of your everyday life. The science capital teaching approach helps teachers to support more pupils from diverse backgrounds to feel able to participate in science lessons. It involves small tweaks to existing teaching and fits with the research messages mentioned earlier (Godec et al., 2017). Ideas from the approach involve broadening what counts as science in and out of lessons, making lessons relevant to their own lives, and linking the science-related knowledge that they bring and that matters to them and their families/communities to curriculum content.

Working with teaching assistants

Research and understanding on how to work with teaching assistants (TAs) has been on something of a journey over the last ten years. A key study from 2009 reported that pupils receiving the most support from TAs made less progress than similar pupils receiving less TA support (Blatchford et al., 2009). Not only was the message that TAs were not aiding learning, but that this was potentially perpetuating the achievement gap as TAs were often working with lower attaining pupils and those with SEND. Research has showed improvement in practice since then, such that the EEF no longer characterise TAs as an intervention that provides negative impact. Targeted use of TAs, where they are trained to deliver an intervention to small groups or individuals, has a higher impact, whereas TAs used as an adjunct in everyday classroom environments has not been shown to have a positive impact on learner outcomes.

In schools you are likely to work with TAs who are either working to individually support a pupil with SEND or are assigned to support teaching and learning in science. However, there ought to be overlap in this and navigating how to work alongside the TAs to maximise the value that they can provide to your science lessons is an important skill. This is something that you should speak to your mentor about and whoever coordinates the work of TAs in your school (often the special educational needs coordinator). Generally, you will want to ensure that you share your plans with TAs, ideally in advance of lessons, although this can be challenging if the TA assigned to a class may vary. You can use lesson plans as a communication tool – share with TAs early and provide a copy of resources at the start of the lesson so they know what is to come and can consider how to support pupils. Ensure that TAs are an addition to your work with students, not a replacement. By that I mean do not see your TA as responsible for teaching an individual or group while you are teaching the remaining pupils. Try to integrate TAs to work with pupils in the whole class, rather than only with a small group or individuals. Ask TAs for feedback about the students they are working with and about your teaching; they will provide a useful perspective.

Help your TAs focus on learning and not on task completion. You want them to scaffold learning activities to help pupils access them, rather than to help the pupils complete a task.

SUMMARY AND KEY POINTS

Adaptive teaching and differentiation in science teaching do not need to involve creating different teaching opportunities for different pupils. It can be thought of holistically to help achieve more inclusive teaching.

Adaptive science teaching is not very different to what we would consider high quality science teaching. Strategies that you might use to help individual pupils in science lessons are likely to be relevant for a range of pupils, perhaps a whole class.

SEND teaching strategies are also not always specialised approaches to make science accessible. Ideas like writing frames and graphic organisers can help pupils with SEND and other pupils by scaffolding concepts in science.

We can also respond to other key issues for inclusive science teaching, drawing upon resources such as the IoP resources for gender-inclusive science teaching, and the Science Capital Teaching Approach, for small tweaks to teaching that can raise participation from pupils from varied backgrounds, as well as helping TAs to focus on supporting learning in science.

FURTHER RESOURCES

https://improvingteaching.co.uk/responsive-teaching/

To go a little further with your thinking about adaptive teaching, you might consider it in relation to planning and assessment. Harry Fletcher-Wood has literally written the book on this, which he calls responsive teaching. In this blog he sets out how responsive teaching can respond to six challenges of teaching.

https://www.wholeschoolsend.org.uk/page/condition-specific-videos

While we do not need to reinvent the wheel and prepare different lessons for pupils with SEND, understanding their individual needs is critical. Whole School SEND, who used to be called SEND Gateway, have a suite of videos on different conditions and inclusive teaching strategies:

https://www.genderaction.co.uk/online-resources

If you are interested in challenging gender (and other inclusion relevant) stereotypes, there are a range of reports and resources from Gender Action.

https://discovery.ucl.ac.uk/id/eprint/10080166/1/the-science-capital-teaching-approach-pack-for-teachers.pdf

Read the Science Capital Teaching Approach report that explains the approach and gives a range of examples of how teaching can be adapted to increase participation.

TEACHING SCIENCE FOR SOCIAL AND ENVIRONMENTAL JUSTICE

Andrew Howes

INTRODUCTION

What do we mean by social and environmental justice, and in what ways are these the concern of the science teacher and science department? These questions can confirm and develop our moral purpose as science teachers. So this chapter presents a challenge: to reflect on our own preconceptions and to enrich our ideas through reading, through activities with young people and with colleagues, and through dialogue.

Social justice can be usefully broken down into issues of representation and distribution (Fraser, 1997). Or to put it more directly – social injustice often involves both a lack of representation as well as unfair distribution.

- *Representation* involves questions such as: whose stories get told (and whose stories are not mentioned)? Whose perspectives are being considered? Who is excluded from textbooks or from classroom discussions? Who gets to be seen and represented? Whose experience counts? Whose understanding matters? There are well-researched issues around these questions in terms of ethnicity, gender, religion, and sexuality, showing how they influence the way young people see themselves in terms of science (Reiss, 2003; Calabrese Barton & Upadhyay, 2010)
- *Distribution*, on the other hand, involves questions of access: who gets access to effective science teachers, to science lessons that feel relevant and meaningful, to community resources, to opportunities for entry to a wide range of careers? This is not just about financial wealth. Socioeconomic disadvantage and poverty are associated with poorer health and damp and crowded housing – but also with lower participation and engagement in science and science-related jobs (Cooper & Berry, 2020).

DOI: 10.4324/9781003110187-20

Environmental justice is closely tied to social justice – so again, it's about representation as well as distribution. Across the world, it is those people and families with access to fewer resources who suffer more from climate and environmental crisis, but their stories are not often told. We face genuinely global crises: increasing temperatures and pollution affect populations of plants and animals, including humans, across the planet, and carbon emissions are tied to global systems of trade and production. But the impacts are felt locally. It is the poorest people on the planet who feel the effects more strongly. That is just as true in Manchester as it is in Mumbai.

Over-reliance on free-market ideas, coupled with inattention to unintended consequences of production and consumption, has led to a planetary crisis manifested in societies across the world (Malm, 2018). But all is not lost – yet. Society is not a collection of individuals with the power to attain and develop just as much as they are inclined. The Covid pandemic has reminded us forcefully that we depend on others but that some people have access to far more resources than others. Society is unequal because the structures of society have stacked up in favour of some groups and against others. But the response to the pandemic has also reminded us that responding to global challenges can restore and strengthen social action in communities. It has also shown the importance of science and scientific thinking for families and communities and for governments and agencies across the world.

OBJECTIVES

At the end of this chapter you will be able to

■ Articulate your commitment to science education as an influence for social and environmental justice
■ Reflect on the hurdles in the way of developing science teaching to realise this influence
■ Draw on examples to help you plan and sustain approaches to social and environmental justice in your own practice as a science teacher.

SCIENCE EDUCATION *CAN BE* A POWERFUL FORCE FOR SOCIAL AND ENVIRONMENTAL JUSTICE

What can science education do? It can unlock, for all young people, an understanding of themselves within a complex and awe-inspiring living planet, within a universe of almost unfathomable energy and time. It can facilitate ways of being in that world as curious and deeply connected people, equipped with tools for thinking critically and analytically and approaching practical and social problems with courage and with an appreciation for the value of collaboration. Or of course, it can switch them off, leave them with the impression that science is not for them and that it has no relevance to the issues that matter to them (Archer et al., 2020).

Science education in societies such as the UK can have a significant influence on young people, where science teachers consider and take account of the following:

- The breadth of the curriculum – there are many opportunities for teaching in the context of climate and environmental crisis, for example
- The reach – science has a large share of curriculum time for all young people, which makes it possible to influence their scientific knowledge, skills, and attitudes
- Highly relevant topics – which can be addressed in science lessons, making links to relevant social action and activism by young people, whatever their particular socio-economic background
- The opportunity to model – and equip young people with – a range of powerful approaches to conceptualising and solving problems at all levels of human experience.

Science teachers can support young people of all backgrounds in using their developing agency in response to significant social and environmental issues. We cannot meaningfully consider the present context in which young people are growing up, let alone their future, without considering environmental justice (Brown & Lock, 2018). The environmental crisis has changed the context of our education system, and science education must play a significant role in response. We need to reappraise what outcomes are of most value, what skills and knowledge are most significant, and how we can support young people to actively address issues of social justice which have a strong scientific dimension.

We will explore the potential of science education for justice shortly. But first, we need to look at some of the reasons why this potential is not being realised.

There are significant barriers to overcome in releasing the potential of science education for justice

Barriers to achieving the potential of science education arise because of contradictions in the systems and practices of education in general and in the wider society. For example, on the one hand we want an inclusive education system in which all young people belong and experience appropriate challenge and achievement – but on the other hand, any such system must fit within the constraints of government budgets. On the one hand we want science education to meet the needs of industry, enterprise, and the knowledge economy for highly skilled young people, accurately labelled through exam grades – but on the other hand we want a curriculum which enables *all* young people to be able to recognise and respond to the value of science in their lives.

Three specific barriers arise from these systemic contradictions: our assumptions as science teachers, a narrow focus on attainment, and systematic bias in our histories of science.

Our assumptions as science teachers

There is a famous test for children and young people, called the 'Draw a Scientist' test (see Chapter 4), which reliably reveals stereotypes of science and scientists held by children: their drawings are often of old, quirky male figures in white coats, with energy rays and boiling test tubes coming out of their pockets, and rats running up their flared trousers. A recent meta-analysis over five decades of use shows persistent stereotypes in many contexts (Miller et al., 2018). However, the problem here is almost certainly deeper than the test suggests. There is evidence that many teachers of science themselves hold quite limited conceptions of science and scientists (El Takach et al., 2020). When science teachers think of science careers, they often consider only professional jobs and research scientists – and this mirrors a narrow view of science (Scantlebury et al., 2007).

Teachers hold other assumptions too, around race and gender but also very commonly in terms of poverty and disadvantage. Most commonly they hold a deficit view, rooted in the unfamiliarity of mostly middle-class teachers with people in poverty. To adapt a question from Gorski (2016, p. 381): Why, on average, do families experiencing poverty not attend science museums with the same frequency as their wealthier peers? Science teachers may easily make unfounded assumptions about how and where science is part of the experience of people in particular socio-economic circumstances.

A narrow focus on attainment

Arguably, in many science classrooms, the focus is on functional science literacy: on the recall of facts and knowledge without much understanding and engagement, sometimes referred to as a 'pedagogy of poverty' (Haberman, 2010). Such a pedagogy supports school performance in a competitive system at the expense of individual or social transformation. A helpful distinction contrasts this with critical science literacy (Calabrese Barton & Upadhyay, 2010) through which young people can challenge normalised and socially understood limits.

A partial, biased, and limited history of science

In many countries, the science curriculum is full of assumptions about whose knowledge counts and which scientists matter (Harding, 1991). This is especially evident in terms of the history of science, dominated by the perspectives of former colonial powers. So there is *silence* on the topic of science and racism and on the role of science in discrimination more widely. Science in the popular imagination – and in the imagined curriculum – remains tied to the myth of human progress (Gray, 2016). But this myth is out of date. Climate and environmental crisis has been brought about by the very human progress which science has facilitated since the industrial revolution. Science can of course still be a force for good, but more than ever we need young people ready to evaluate the application of science. Arguably, concepts such as scientific literacy contain very little that is critical of the role of science, either historically or in the present. The assumption in the science curriculum is that science is a good thing, and that is simplistic and lacking in credibility.

The underlying contradictions cannot be easily dealt with, and there are no easy solutions. But there can be better resolutions of the contradictions, as we seek to overcome some of the key barriers.

SCIENCE TEACHERS CAN BE AGENTS OF CHANGE

Science education takes place in the context of huge assumptions about the young people we teach and the effect that our science lessons have. Lemke (2001, p. 305) sums this up very neatly:

> We have imagined that the few minutes of the science lesson somehow create an isolated . . . learning universe, ignoring the sociocultural reality that pupils' beliefs, attitudes, values, and personal identities – all of which are critical to their achievement in science learning – are formed along trajectories that pass only briefly though our classes.
>
> (Lemke, 2001, p. 305)

Science teachers have a limited influence: the question then is how we can use this wisely.

In this section, we look at some approaches that some science teachers are already working on, addressing social and environmental justice through science teaching. Throughout, we reiterate an underlying theme: that getting to know your pupils helps to reduce barriers.

■ Questioning our stereotypes and assumptions
■ Representation and recognition
■ Connectionist science teaching
■ Challenging discourses of poverty in and through science education
■ Pupil voice in science education
■ Science education for climate crisis
■ Tackling conspiracy theories

These approaches overlap to some extent – they are different spotlights on the same question, ways that science teachers can make a difference through their teaching.

Questioning our stereotypes and assumptions

A powerful place to begin is with our own stereotypes and assumptions. Of course, it is the nature of assumptions that they are not easy to notice. And yet there is evidence that we are likely to hold many relevant and limiting stereotypes, on, for example, science; scientists; young people in poverty; young people as activists; disadvantage; sexism in the classroom; racism in the curriculum;

environmental crisis; the role of science education; the limitations on science teachers; and *who* has relevant knowledge and experience to bring to our science lessons (Kahneman, 2011).

So as a starting point, and as an example of the sorts of questions which might help you to unpack your assumptions about other relevant issues and groups, Task 19.1 asks you to consider your own stereotypes about science and scientists.

Task 19.1 **Who does science?**

The idea that science is done by scientists is very limiting. Ask yourself the following questions: *which jobs* do you think require some scientific thinking or understanding? What do your pupils think? *Where* do people do science? Only indoors or only in laboratories? What do your colleagues say about this? And when we say science, *what* science are we talking about? Science is a such a broad label – so more specifically, *who* might know *what* about medicines, or engineering, or technical jobs, or crafts, or alternative fuels? Or about the strength of textiles, the taste of cake, the physics of games, diagrams of heating systems, rewilding in local communities, the chemistry of water, the microstructure of bone? Over the next week, have a conversation with someone new about science, and use it to construct a lesson activity.

Representation and recognition

Biology is rich with powerful concepts, such as ecological diversity and interdependence. In socio-cultural terms, both diversity and interdependence have come to stand for important dimensions of social justice and can help science teachers to recognise the richness of their classrooms, the communities around the school, and the many different, interconnected, and *equally valuable ways of life* which people lead. Task 19.2 is about building an appreciation of that richness.

Task 19.2 **Describe the diversity of your class in a new way**

We often think in terms of biological characteristics like height, skin colour, mass, lung capacity, resting pulse – some of which are inherited, others environmental, or both. We have learnt to respond to categories such as pupil premium, special educational needs, and EAL (and the many differences those categories often disguise). But what if we as science teachers showed more interest in pupils' interests, their life histories, neighbourhoods, family stories, experiences of illness, the challenges that they have faced? Certainly this needs to be done with care, as well

---

as maintaining a secure classroom environment where no one feels judged. But getting to know pupils better is a way of humanising your classroom. As you move around the school and watch experienced teachers in action, think about how they get to know their pupils as individuals.

Next, think about how can you increase your knowledge of the diversity of your class and their knowledge of each other, as part of your science lessons? For example, you could ask pupils to write or talk about someone they know who has a job using science in some way. Or ask for a volunteer willing to demonstrate some techniques from their favourite sport, then link these to science. Or you might create a participative map of air pollution hotspots in the neighbourhood – or eco-diversity hotspots – and discuss the impacts that they have noticed.

Write a plan for a short section of a lesson in which you deliberately create links between the scientific content and pupils' interests or local environment.

A critical further step is for your taught science curriculum to explore socio-cultural diversity and interdependence in the history and practice of science. In developing schemes of work (see Chapter 8), it can be helpful to think in terms of *past, present*, and *future*, for widening representation in the science classroom.

In terms of the past, it is now not difficult to ensure that we go beyond the dominant, narrow, Western account. Our histories of science can easily make reference to figures in chemistry, medicine, astronomy, and optics from Arabic science, during the 'Golden Age'. The connections are there in many fields: in language such as *al*cohol; in the systematic advances in classifying chemicals; and in advances in scientific methods. We can show links to Arabic conceptual developments in understanding the cardiovascular and nervous systems and in understanding human beings as evolving from animals, as well as the nature of light and vision. Work is also being done on the history of science and technology developed across other civilisations and countries, including astronomy, metallurgy, and medicine across West and East Africa for example. Science has never been less than a multicultural human endeavour.

In the present, there are endless examples of contributions to science from across the world, through international collaboration, and countless topical links to make. Turkish-German scientists (Sahin and Tureci) created the first Covid vaccine available for public use.

Regarding the future: recent work on the invisibility of women in the design of products, systems, setting of British Standards, etc. (Criado Pèrez, 2019) further strengthens arguments for more women in science and engineering, noting how, for example, standards set for seat belts make collisions safer for the average man but dangerous for the average woman. Studies have explored the roots of under-representation in ethnic minority groups (Archer et al., 2015b).

Science education should be a curriculum subject where explicit appreciation of contributions and participation by people of diverse ethnic background, gender, and sexuality is normal and evident in our lessons. We can commit to unpacking these ideas within science lessons. Representation helps. But it is at least as important that science teachers check our language and assumptions in relation to under-represented groups in science and facilitate critical conversations about the consequences of this under-representation in so many fields. In terms of careers, we can choose diverse examples of people using science in their jobs. In Task 19.3, you will put these ideas into practice in a concrete example by designing a lesson.

Task 19.3 **Design a lesson which focuses on the diversity of people working in science today**

Using whatever lesson planning format your Initial Teacher Education (ITE) provider or school uses, design a lesson with the objective of helping pupils understand the diversity of people who work in science or use science in their work. For example, you could set up a research task for pupils to create videos introducing scientific work on a particular topic by women and men from different backgrounds.

The next section describes an even more direct approach to building on the diverse knowledge and experience of the pupils in your class.

Connectionist science teaching

Another way to develop recognition is to teach very explicitly from what young people already know about – and which you as a teacher may *not* know about. Most 'real world' examples linking science to the everyday are chosen and developed by teachers and curriculum writers. In contrast, connectionist science teachers seek out examples from the real world of the young people so that lessons are grounded in young people's experiences in life: their contextualised and gendered agency in family and community contexts and their encounters with science through the media, leisure, health, housing, work, transport, etc. It is informed by theoretical and methodological traditions including the funds of knowledge work by Gonzalez et al. (2002) and the experience of engaging young people in research, such as participatory photography (Miles & Howes, 2015).

Connectionist science contrasts with the 'science capital' approach (Archer et al., 2012), particularly where that capital is assumed to depend on visits to science museums, for example or having a family member who is a doctor or engineer. Learning first what young people know about and are fascinated by might lead to rich learning experiences built around local amenities such as shops, pharmacies, buses and bikes, or patches of grass. But finding out from young people comes first. This approach is especially helpful as a way to challenge the deficit discourse around poverty. This discourse is pervasive in many schools, and we look next at how we might challenge it through science education.

Challenging deficit discourses of poverty in and through science education.

Research has shown how hard marginalised pupils have to work to make meanings about science. For many young people success in the science classroom requires border crossing between subcultures (Aikenhead, 1996). But remember: social injustice involves unfair *distribution* as well as lack of *representation*. Limited and expensive public transport; parents working long hours in low-paid jobs; and difficulty in finding spare cash for school trips all influence young people's opportunities. These are the starting points of a much more powerful *structural* explanation for the relationship between poverty and education (Gorski, 2016), challenging a dominant view in which learners, parents, teachers, and schools are labelled as 'failing' and, in effect, *blamed* for the stubborn link between poverty and poor academic achievement. This blaming is highly problematic, because it suggests that the teacher's role is to make up for the perceived deficit, often by adopting a teacher-centred focus on facts and recall (Haberman, 2010). In contrast, research recognises the power of enquiry-based science lessons to more fully engage pupils (Thadani et al., 2010).

Stereotypes of people living in poverty continue to influence responses to poverty in schools, but as science teachers we can actively refute those stereotypes and help pupils to recognise and consider the structural causes of poverty. A very significant challenge, which science teachers can directly address, concerns health outcomes. Work on the social determinants of health is supported by rigorous epidemiological evidence (Marmot, 2020). Health outcomes, which are so often presented as the responsibility of each individual, in fact depend hugely on class, race, and gender, because of different levels of access to material resources such as housing, conditions of employment, transport, and food security (Pickett & Wilkinson, 2010). During the pandemic, the influence of these social inequalities was evident in the variations in serious illness and death (Paremoer et al., 2021). As science teachers, we have a role in promoting critical thinking, to test and disrupt established discourses about poverty.

Connectionist science is another way forward, if we commit to learning from young people and building curriculum links from what we learn. For example (Beckett & Wrigley, 2014) list a positively framed set of questions that teachers could ask in class, some of them potentially useful for science teachers, such as:

■　Who do you know who's got an interesting job?
■　Do you know anybody who grows things?
■　Have you travelled to any interesting places? Who do you know who has?

So having considered connectionist science approaches and the need to learn from pupils, Task 19.1 asks you to plan some questions like these but specific to the science topics you are going to be teaching in the near future.

Task 19.4 **Plan some open questions to ask pupils**

Plan part of a lesson in which you ask pupils questions about topics that will give you insight into their knowledge and experience and which you could link to in class. Notice that these are all questions that you as the teacher do *not* know the answer to. For example:

■ Do you know anyone who has to wash their hands at work? Do you know why they have to do that? Can you ask them and tell us next lesson?
■ Who do you know who watches loads of nature programmes? What do they enjoy about them?
■ Does anyone know the word for 'water' in a language other than English? And when you say that word, what else comes to mind?
■ Do you know anyone who uses chemical solutions as part of their job? Can you find out what it is used for and what it is called?

Use the pupils' responses to these questions to link into teaching your current topic.

Implicit in many of the approaches to science education for social justice is that teachers make time to listen to pupils. The movement towards pupil voice has a lot to offer in science education, and the next section contains a vivid example of that.

Pupil voice in science education

There is extensive evidence for empowerment and valuing of science through pupil voice in science education (Laux, 2018). However, this can be difficult to put into practice. Paying attention to pupil voice involves listening, with teachers first trusting in students' ability to hold or express a valid opinion (Messiou & Hope, 2015), countering the 'ideology of immaturity' (Rudduck & Fielding, 2006). Second, teachers need to create opportunities for young people to express themselves, and third, they need to listen.

Albalawi (2021) describes a powerful example of pupil voice. In her research, she used photovoice methodology (Miles & Howes, 2015) to invite young people to take photographs of their home or locality in response to prompts about the relationship between science and the real world, then listened to the young people's commentaries on the photos and their meaning. A year 9 student took a photograph of the ground (Figure 19.1) as something which inspired her to do science. Albalawi notes,

> Ayah (pseudonym) was curious to know what the ground is made of, when it was made and what consistency it has. She assumed that we ignore or, in other words, do not pay much attention to the ground because it is something we see as insignificant . . . Ayah sees science as positive, which made me want to know what science means to her. Ayah views science as

something that can explain the world . . . it is the process of enquiry. Ayah then moved on to discuss what can be a sensitive issue in the Muslim world: different explanations of the world, based in religion and in science. Ayah is a hijab-wearing Muslim and talked about how Muslim people may understand science, and how some religious people do not believe in science – and gave her own thoughts on this.

(Albalawi, 2021)

This example shows how much teachers can learn about pupils' ideas through a simple activity, helping us understand pupils' ideas and models of the world and their diverse and distinctive thoughts about science *in* the world. Whilst we might not have curriculum time to elicit pupil voice in the detail described in this

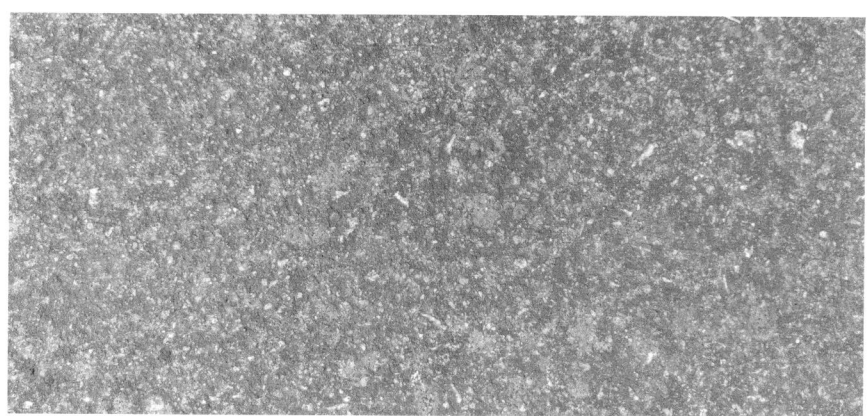

The ground. It has inspired me because [it] is like something [that] is really small. We look to the sun, we look to the plants, but we don't really notice [what] the ground is really made of, so I think [if] we know [what] this stuff made of, I think that's really fascinating and that's why I want [to learn] science and how this was made and when [it] was made.

Do you think the thing that made the ground is related to science?

Yeah.

Do you think you [are] going to do in [the] future anything related to science?

I wanted to do something… I don't know what is called, it has something to do with geography and science, people who go on mountains.

What is science for you?

For me, I think like explained the world to me. For example, religion and science [are] completely different things, but even though, like, I think some [religious] people, they don't believe in science, but I think [they] go hand in hand, beside in wonder. [If] you are a religious person, you [are] going to see the world in [a] scientific way and the logic behind everything is science.

■ **Figure 19.1** An example of photovoice to elicit pupils' ideas about science (Albalawi, 2021)

research, it is worth considering how pupil voice can be drawn on as part of our science teaching.

In the concluding section of this part of the chapter, we turn to the environmental crisis, which is arguably the most urgent social justice issue for our time. To address the crisis as science teachers, we need to bring together all the possibilities that we have discussed for working as agents of change. The crisis demands that we reflect on our intentions for the curriculum in the light of social justice. It also requires of us that we deepen our understanding of young people – who they are, what they want, and what they offer.

SCIENCE EDUCATION FOR CLIMATE AND ENVIRONMENTAL CRISIS

The climate and environmental crisis is a crisis of social justice. Young people perceive this clearly, not least because the science of climate change provides clear messages about adverse impacts on current and successive generations of human beings. This clarity on the part of young people potentially marks a sea-change in their relationship with science teachers. Many young people recognise that science matters hugely, yet science pedagogy and the science curriculum lag behind (Rousell & Cutter-Mackenzie-Knowles, 2019). Young people are building organisations (like Teach the Future) to press for change. Science teachers need not only to listen but to learn to respond to pupils as activists and to work with teachers in other subjects to support their actions. In the short term, we need to make spaces in our curricula, while expanding our pedagogical imagination.

The slogan 'Think Global, Act Local' has gained traction with awareness of the climate crisis. Local responses include learning walks around the school to map wasted energy; participatory photography and video to engage with the perspectives of young people and their parents; participatory mapping in relation to biodiversity in and around the school grounds. Teachers and pupils can use these approaches to develop local projects – which might focus on resilience in school grounds – or biodiversity, rewilding, and the powerful educational concept of nature connectedness.

Task 19.5 An urgent, local, educational response to the climate crisis

Anderson et al. (2020) argues that an equitable allocation of the world's carbon budget means that advanced economies have pretty much used up their share already and must accept the fastest transition to renewables. Do you agree? What are the implications for the UK in terms of heating, transport, industry, and food? And do you recognise any implications for your teaching of science?

Either on your own or in collaboration with another student teacher in geography or science, plan a lesson or extra-curricular session that explores these questions with pupils and relate it to the science curriculum. If possible, invite a local individual or organisation to help you and your pupils to address an aspect of climate crisis.

Finally, I argue that science education should challenge the purveyors of selective truth. For example, there is compelling evidence that fossil fuel businesses have lobbied successfully over decades to avoid a focus on climate change, much like the tobacco lobby decades earlier. Critical science literacy involves developing both awareness of people and institutions who might want to manipulate scientific evidence for their own benefit and the ability to critique claims scientifically. Science teaching for social and environmental justice should aim at supporting young people to notice conspiracy theories and be skilled at evaluating significant claims in relation to health, environment, and social justice (Lewandowsky & Cook, 2020).

SCIENCE TEACHERS CAN GET SUPPORT

Whether we start from environmental justice or social justice, the possibilities of change in and through science education are huge. This is both exciting and potentially overwhelming. Happily, once you start to look for them, there are many potential partners, networks, and groups who are keen to support you in this work as a science teacher. But often, teachers forget, or don't have time, to look.

As a teacher in a school, you quickly get used to being part of the institution, working with your colleagues and with your classes, day to day, week to week, term to term. It is often hectic. Quickly you become used to being an insider, and you lose awareness of the astonishing and direct access that you have, within the school, to working systematically and personally with so many different young people. *Meanwhile*, on the outside, there are large numbers of organisations who want to support teachers, either directly or through resources and networks. But they find it very difficult to get access to schools. Schools can be very protective and inward-looking institutions, although at some level they aspire to connections beyond their fences and gates.

This puts individual teachers in a very influential position – as potential brokers of connections and partnerships which might well enrich the learning of young people. Make some time to search for partners, make connections, and draw on the help that is available to you in developing science teaching for social and environmental justice.

SUMMARY AND KEY POINTS

This chapter has argued that science education has a huge (but often untapped) potential in relation to social and environmental justice. It has suggested that there are significant barriers in the way, including the narrow focus on attainment scores, with some potentially negative pedagogical consequences in a competitive school system. Some of the barriers are due to habits and norms within science education itself, such as the examples that we choose to focus on and the questions that we do and do not ask. Nevertheless there are many approaches that can be taken by teachers who want to construct a more relevant, engaging, and responsive approach to science education, with the agency of teachers and young people at its heart. These include methodologies for teachers to learn about science in the everyday lives of young people in the class; building a critical awareness of assumptions about poverty, race, and gender, especially in the science curriculum and in the science classroom. Such approaches are more engaging for young people and for teachers and help to dismantle significant barriers to participation in science. Such approaches are also likely to lead towards social and environmental justice, within the science classroom and also outside and beyond the classroom. And happily, there are many partners outside schools who are keen to work with science teachers. You do not need to struggle alone.

Check which requirements for your initial teacher education you have addressed through this chapter.

FURTHER RESOURCES

Calabrese Barton, A., & Tan, E. (2019). Designing for rightful presence in STEM: The role of making present practices. *Journal of the Learning Sciences*, *28*(4-5), 616-658.

A powerful case study of pedagogical design for recognition and the 'rightful presence' in classrooms of people from minority backgrounds.

Haberman, M. (2010). The pedagogy of poverty versus good teaching. *Phi Delta Kappan*, *92*(2), 81-87.

This remains a fresh critique of disengaging lessons in a high-stakes system.

https://www.campaigncc.org/schoolresources compiles a wide range of resources for enriching the school curriculum to better address the challenge of climate and environmental science.

Lemke, J. L. (2001). Articulating communities: Sociocultural perspectives on science education. *Journal of Research in Science Teaching*, *38*(3), 296-316.

A key text that repays repeated readings.

Lewandowsky, S., & Cook, J. (2020). *The conspiracy theory handbook*. http://sks.to/conspiracy

A compelling way of framing the widespread need for scientific thinking.

Perez, C. C. (2019). *Invisible women: Exposing data bias in a world designed for men*. Random House.

A powerful feminist critique of the false assumptions of objectivity in scientific data that relate to human systems and choices.

IS EDUCATION RESEARCH VALUABLE FOR TEACHERS OF SCIENCE?

Lee Elliot Major

INTRODUCTION

In this chapter I will outline the main findings from a synthesis of education research studies highlighting the approaches that are most likely to improve learning in the classroom. My central argument is that science teachers should aim to be evidence-informed in their practice but also be aware of the limitations of education evidence and how it differs from scientific inquiry in the natural sciences. The chapter will offer some practical suggestions for reflecting on evidence and research to help you develop your own classroom practice.

Taking general research findings and making them relevant to your own science teaching is a challenge that you will have to grapple with as an educator. How can research help you improve the learning of your pupils? Can it help you make your teaching more effective or efficient? As a teacher of science, you can reflect on the similarities but also differences in how research is undertaken in the natural and social sciences.

OBJECTIVES

At the end of this chapter you will be able to:

- List the best bets for improving learning and the education myths with little evidence of impact
- Know the differences between evidence gathering in education and scientific inquiry in the natural sciences
- Understand the uses and limitations of education research in helping to improve your science teaching practice
- Consider practical ways of how to utilise evidence to improve your own teaching.

DOI: 10.4324/9781003110187-21

WHAT WORKS?

'Learning the facts about learning' ran the BBC headline for a news article on the UK government's new 'What Works' initiative in 2014 (Easton, 2014). A series of sub-headings declared apparently unassailable truths from education research: 'Setting and streaming? NO. Teaching assistants? NO. Feedback? YES. Peer-tutoring? YES'. The drive to create 'evidence based' policy and spend public money on the most effective practices was a welcome development in England. And the synthesis of education research that I had helped to create was championed as an exemplar for the UK government's new network of 'What Works' centres.

But the nuances, caveats, and ambiguities of our research summaries had got lost in translation. Carefully worded suggestions for the best bets for improving learning had been reduced to crude certainties. Classrooms of pupils were assumed to be the same homogeneous units across the world. The complexities of converting research into practical advice had been bastardised into a binary code.

The experience highlighted the power but also dangers of promoting a more 'scientific' approach to classroom practice. Education research can only give us an indication of what has worked in the past, not what will work in the future. Indeed, it can only offer indications of what may work under certain conditions. Science teaches us that we should tread carefully before reaching any conclusions.

SUMMARIES OF EDUCATION RESEARCH

My hope in co-writing the 2019 book 'What Works? Research and evidence for successful teaching' (Elliot Major & Higgins, 2019) was to encourage teachers to embrace the sceptical and constructive mindset of the scientist: challenging pre-conceptions and lazy assumptions over what has worked in the classroom but also embracing the inherent uncertainties of any professional practice. Given limited budgets and time, it is critical to reflect on what impact we're having. The book provides best bets for learning based on the aggregated results of over 200 summaries of 8,000 intervention studies. Our advice covers a range of teaching approaches from delivering classroom feedback to one-to-one tutoring.

The book built on a decade of work developing accessible research summaries for teachers. My co-author, Steve Higgins, is an expert in the use meta-analysis, an approach which combines hundreds of studies from across the world to produce overall findings. This provides the overall view, rising above the contradictory and conflicting conclusions from single studies (Higgins, 2018).

Our first toolkit for teachers (Higgins et al., 2014) was a 20-page report for the Sutton Trust charity presenting evidence summaries for 20 teaching approaches. We wanted to help teachers deploy their pupil premium funds for poorer pupils effectively. This evolved into what is now known as the Education Endowment Foundation Teaching and Learning Toolkit, a 30-strand interactive website you can freely access. By 2015 two-thirds of school leaders across

England reported they had used the toolkit (Cockburn et al., 2015). Versions have been launched in Scotland, Australia, Spain, and Latin America.

We thought hard about the design and presentation of our findings. These include the costs of implementing approaches and the strength of evidence underpinning our conclusions of how well it had worked. Most importantly, we translated effect sizes, the comparative impacts of interventions, from the language of statistics into a more meaningful measure for everyday classroom use: extra months gained during an academic year. Without this, our summaries would not have caught the imagination of busy classroom professionals.

Context of evidence summaries

The take-up of these guides highlights the appetite for evidence among teachers as they seek to improve their practice. But there are some important caveats you need to bear in mind before you jump to conclusions that research can instantaneously transform your science teaching.

First, learning to teach is complex. You are trying to organise around 30 pupils while undertaking activities enabling them each to gain new skills, knowledge, or understanding, taking into account the short-term and long-term effects of your actions, considering where each pupil is on their learning journey. The highly complex and constantly changing human interaction of the daily classroom contrasts with the controlled experimental conditions of natural sciences where universal immutable laws appear to guide physical or natural phenomena. Science teachers can find the transition from laboratory or field-based scientific study in their undergraduate years to the less certain paradigm of social science in education challenging.

Beware research findings from other fields that have not been applied in classroom and curriculum contexts. Psychology research on motivation or neuroscience research into brain function can pique the educator's interest. But how applicable are they to the real world of the classroom?

Education evidence can provide useful tips for you, but many of the decisions you make will come down to your professional judgement at a specific moment in time in a certain classroom context.

Second, the studies summarised in our guides adhere mostly to the 'positivist' approach in the social sciences. Positivists adopt the same methods to study the social world that scientists use to investigate the physical world, working under the assumption they can identify universal patterns that shape our actions. Other researchers advocate an 'interpretivist' approach, believing that scientific methods are not appropriate for the unique web of interactions between teachers and pupils and focusing on the subjective experiences of individuals.

Our work adopts a pragmatic middle-way: aggregating the impacts of scientific trials on the one hand but emphasising that these only offer best bets to aid professional judgement in a particular context. Another difference between education research and scientific inquiry is the absence of replication studies in education. There are relatively few trials that evaluate the effectiveness of approaches in a range of school contexts. You should be extremely cautious about doing anything on the basis of single studies.

Third, you as a science teacher can only do so much in influencing the outcomes of your pupils. Disentangling family and school influences in studies is extremely difficult, and teachers are concerned with a range of positive pupil outcomes – self-esteem and well-being for example – not just test scores. But most studies estimate that the majority (around 70–80 per cent) of the variance in test scores is explained by individual background factors with a much smaller proportion explained by school factors (Hanushek, 2016). This is hardly surprising given that children spend most of their lives outside the classroom. But this inconvenient fact has not prevented governments from placing increasing expectations on teachers to address society's stark inequalities.

Task 20.1 **Contrasting scientific inquiry in education and the natural sciences**

With another student teacher, list the similarities and differences between scientific inquiry in the natural sciences and evidence gathering in the field of education. Do you think the discipline of teaching and learning will become more 'scientific' as more trials are undertaken in schools? Will there always be limits to identifying the same patterns of activity in different classrooms at different times and in different contexts?

Task 20.2 **What factors impact pupil outcomes aside from teaching?**

Should teachers be held to account for all the outcomes of their pupils? Reflect, with a mentor in your placement school, on the evidence detailing the factors that impact on pupils both inside and outside school. Do you think that current school accountability measures are fair?

FINDINGS

The overarching message of our What Works? book will not surprise experienced teachers. The following graph presents the approaches in terms of their average impact and their estimated cost (Elliot Major & Higgins, 2019). What matters most is the classroom interaction between teacher and pupil: providing and receiving effective feedback that moves learning on; encouraging independent learning through metacognitive (or thinking about thinking) strategies and effective classroom teaching in general.

Reducing class sizes has surprisingly limited impact on pupils. Smaller classes work when teachers change the way they teach, catering to individual needs of pupils and receiving more feedback from children. It is not reducing class size that matters but how you adapt teaching style with fewer pupils.

Interpret these headlines carefully. It is true that poorly managed and prepared teaching assistants have little impact on learning (see chapter 18). But that's the whole point: they need to be managed by teachers and be prepared and trained. Teaching assistants can be invaluable secondary educators (Webster et al., 2015).

Task 20.3 **Reviewing findings about what works**

With a fellow student teacher, review the results presented in Figure 20.1. What surprises you? To what extent do you think the findings would change in another ten years when more studies are available? Discuss the different interpretations that can be made from headline research findings; how could we avoid misinterpretations? Discuss with your subject mentor what are the implications for your science teaching.

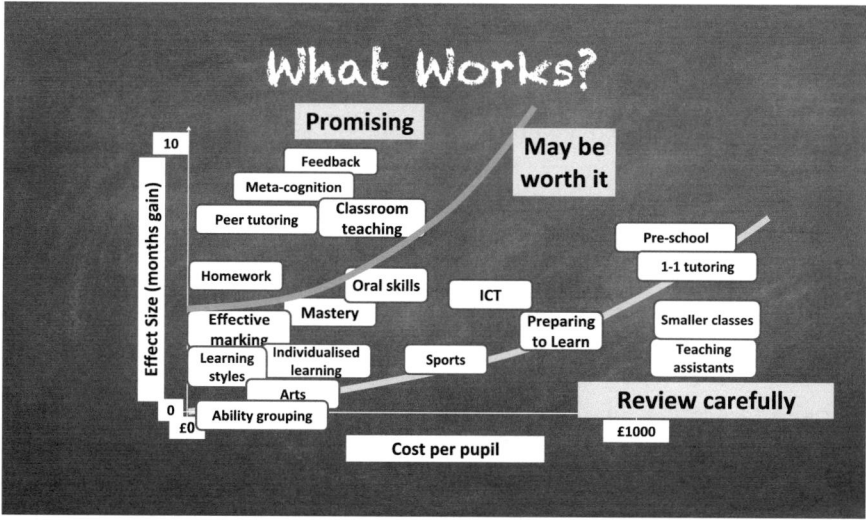

■ **Figure 20.1** What has worked: the effectiveness and costs of different school approaches and activities

Source: Elliot Major & Higgins, 2019

Effective feedback

There is a catch with these findings: the more you focus in on classroom interaction, the harder it is to come up with simple formulae detailing the necessary steps to take. There is a wide variation in impact for a given approach. It's not what you do but how you do it that counts. If delivered poorly it can harm pupils' progress.

Providing specific, timely, and focussed feedback in the classroom is the most impactful approach, boosting the learning of pupils by an extra eight months during an academic year. Recent analysis suggests more realistic gains of three to four months are achievable (Wisniewski, Zierer & Hattie, 2020). However, feedback interventions can also slow down learners. It is how comments are delivered, received, and acted upon that matters (see chapter 16).

We define feedback as information given to the learner and/or the teacher about the learner's performance relative to the learning goals, which can then redirect the teacher's and the learner's actions to achieve the goal. It is the crucial second step in teaching after initial instruction. It can be verbal, written, or given through tests.

Effective feedback makes learning more efficient by changing what science teachers and learners do next. It speeds up the learning process by identifying what the learner needs to do differently to succeed. But it's hard to do. It is easy to fall into the trap of thinking that a few questions answered by a few pupils signals general understanding of the whole class.

In his book *The Hidden Lives of Learners*, Graham Nuthall (2007) revealed the challenge every classroom teacher is up against: 70 per cent of pupils' time was spent pretending to listen; 80 per cent of the feedback pupils' received was from their fellow peers; and 80 per cent of this was wrong!

It is important for you to understand that feedback is not marking. Verbal feedback is usually more effective. Marking is only feedback when it changes what learners do next. Marking has become an unhelpful burden for many science teachers. This is usually because it is less about pupil learning, and more about demonstrating to parents that teachers are busy. It only helps when pupils (or teachers) do something different as a result.

Effective feedback has been shown to work in science teaching specifically. An analysis of results from the OECD's Programme for International Student Assessment (PISA) based on data from more than 540 000 15-year-olds in 72 countries found a positive link between 'adaptive teaching' in science lessons and pupils' science achievement (Mostafa et al., 2018). This appears to reinforce the point that feedback is a two way process between pupil and science teacher and is only powerful if acted upon, not just given. As discussed in Chapter 19, adaptive teaching involves teachers adapting the structure and content of their lesson in light of the knowledge, skills, and interests of their learners. Alarmingly, as much as 27 per cent of pupils reported that their science teacher never or almost never tells them how they are performing.

Homework

Homework, if managed and monitored well for secondary school pupils, meanwhile can lead to five months' extra progress over one academic year. But in the absence of true experimental studies, this research area suffers from the chicken and egg problem. Studies show schools that give more homework perform better – but can't rule out the possibility that their better results could be due to something else the schools are doing rather than the homework itself. Does homework lead to better results or are high-achieving pupils simply doing more homework?

Homework is a risky teaching strategy. There is little control over who completes the set tasks and plenty of potential distractions. The most effective homework has clear aims, is short and focussed, and complements classroom work. You should reinforce activities covered previously or preview topics to come. One study found homework graded or commented on by teachers had twice the impact of homework without any feedback. Homework has been found to be particularly effective in mathematics and science (Canadian Council on Learning, 2009).

Task 20.4 **Applying evidence summaries to homework practice**

Review the summaries on both homework and feedback in the Education Endowment Foundation Teaching and Learning Toolkit (or our *What Works?* book if you have access to it) in more detail and use this to reflect on a homework exercise you have run or observed in school. Discuss with your mentor how you would make any changes to the homework exercise process in light of your reading. Discuss with another student teacher how you might improve the effectiveness of your practice in light of the evidence on homework and feedback more generally.

CLASSROOM MYTHS

While education research tends to focus on 'what has worked' in schools, it is equally important to know 'what hasn't worked'. Myths, traditions, and fads endure in education despite little evidence of impact on children. The most dangerous classroom myths are those that have the scientific ring of truth to them. They often make intuitive sense.

Learning styles

The widespread belief that pupils can be classified as visual, auditory, or kinaesthetic learners, for example, is persistent, despite several reviews debunking 'learning styles'. The idea behind learning styles is that as individuals we all have a particular approach or preferred 'style of learning'. According to this approach, pupils categorised as having a 'listening' learning style could be taught more through storytelling and less through traditional written exercises. One survey found over nine in ten teachers agreed with the claim that individuals learn better when they receive information in their preferred learning style (Howard-Jones, 2014).

Yet the evidence is clear: there are no benefits from the approach. One study found no link between pupils' learning-style preferences (visual or auditory) and their performance on reading or listening comprehension tests (Husmann & O'Loughlin, 2018). Another study found no evidence that pupils who said they preferred learning visually remembered pictures more than other pupils, while those who said they preferred learning verbally were no better at remembering words better (Knoll et al., 2017).

What the research does show is that all learners can think in words and in mental images, and different classroom tasks require different modes of learning. For example, visualisation is particularly valuable for some areas of mathematics. But it is unhelpful to assign learners to a supposed learning style. You should vary the way you present information and encourage learners to understand their strengths and capabilities, but do not limit learners by targeting what you think is their 'style'.

Setting

Grouping children into sets according to their current performance, meanwhile, makes little difference to learning outcomes – and this is the case for the sciences and maths, where setting is common, as in other subjects. In theory, it allows science teachers to target a narrower range of pace and content during lessons. But in practice it creates an exaggerated sense that pupils are at the same level. The gains seen for higher achievers flourishing in the top sets is offset by the damage done to pupils languishing in the bottom classes (Francis et al., 2017).

Schools also do not actually test for ability; they group on current performance. Otherwise, you wouldn't have so many summer-born children in what are called 'low-ability' sets. It is easy to slip into a fixed mindset mentality. Rigid ability grouping can crush the self-belief of pupils who feel no amount of hard work will allow them to escape from the bottom of a class. Pupils are acutely aware of any ceilings imposed on their learning.

Consider carefully your science teaching when school placements are using sets in science. If they do, monitor the progress of children. Consider the pace of learning in lower sets and how pupils can catch up on parts of the curriculum they miss. Watch out for pupils struggling in higher sets – the small fishes in big ponds.

LIMITATIONS

As you are at the beginning of your teaching career, now is the time to embrace an evidence-informed approach to your practice. Remember this is about being open to new ideas as well as retaining a healthy dose of scepticism. It's hard to change habits ingrained over a long career – even when we're presented with evidence that suggests we should change our approach. Consider how this relates to barriers to scientific progress discussed in Chapter 19.

Ironically, there is scant research into teachers' use of research. But what we do know points to little change in behaviour in the light of new evidence. Surveys of teachers have revealed greater recognition of evidence but few signs of more focus on the approaches shown to offer the highest chance of improving attainment. Efforts to disseminate research meanwhile – from research champions in schools to research learning communities to evidence informed professional development – have as yet produced no evidence that teachers were more likely to use evidence or indeed that outcomes of pupils were improved.

Education trials are characterised by 'super-realisation', with researchers providing unrealistic assistance in the delivery of the approaches. Once scaled up, it is impossible to replicate these conditions. Teachers are seldomly involved at the outset of research trials.

The education world is also an evolving ecosystem, where what works changes over time. We have improved our evidence on teaching assistants, for example, and we know that, when properly trained, they can boost pupils' learning by up to five extra months in an academic year (Sharples et al., 2018).

Improving outcomes for disadvantaged pupils

My work is focused on improving the outcomes of disadvantaged pupils in particular. Our challenge is the tendency for the divide between the education haves and have-nots to widen in the classroom. It is termed the Matthew Effect after the biblical reference: 'For unto everyone that hath shall be given, and he shall have abundance; but from him that hath not shall be taken away even that which he hath' (Matthew, XXV). Pupils who fall behind in reading read less. Poor reading skills then inhibit their learning in other subjects. Disadvantaged pupils are more likely to be selected into lower sets.

You will need to be acutely aware of the myriad ways in which disadvantage manifests itself among children. When you are born, as well as where and how you are born, has a profound impact on your education prospects. Children born between June and August have lower self-esteem, are less confident in their own abilities, and are more prone to fall into risky behaviour (Crawford et al., 2013). If we aspire to have an equitable school system, then we must address summer-born disadvantage.

As far as we can tell there is not a particular pedagogy of poverty – approaches that have extra benefits for disadvantaged pupils. Effective teaching benefits all children, including those from less and more supportive home environments. Metacognitive approaches that make learning goals explicit often help lower-attaining pupils more than their classmates who are already successful. Another approach is to enable children to be better prepared for learning – providing free breakfasts, offering sleep education, or helping to get glasses for poor eyesight.

Task 20.5 **Reviewing research to draw implications for your developing practice**

Find two other student teachers or an early career teacher in your school (they do not need to be teaching science) and choose a topic with the aim of improving one area of practice. Review the relevant area of education research, using summaries of academic literature from one of the resources listed at the end of this chapter. These will also include references to individual research papers. Are there any surprises in the findings? What are the limitations of the research? Is there enough evidence to inform your practice? Where appropriate, explore implementing one of these evidence-informed approaches in your practice, then meet again to talk about how it went.

Improving praise

Praising pupils can feel like the right thing to do – affirming the work of your pupils. But studies suggest the wrong kinds of praise can do more harm than good. Praise meant to encourage low-attaining pupils actually conveys a message of low expectations. Criticism of poor performance can indicate a teacher's high expectations. Praise is valued more when it is meaningful, focussed, and scarce.

Avoid personal feedback – such as 'you are clever' – it provides no information for moving learning on. Praise should be directed to increasing pupils' efforts to understand a problem: 'You've done well because you have completed this task by applying this concept'. It is worth spending some time thinking about how you can provide specific, positive feedback that moves learning forward, rather than generic praise.

Focusing on your hidden learners

Think of ways to elicit feedback from every pupil in your class to know where they are in their learning. Too often teachers rely on a select number of responses from a few dominant voices in the classroom to gauge whether learning has occurred. Involve the quieter and more vulnerable pupils. It's these 'hidden learners' who may be pretending to listen!

Avoid simply asking pupils to put their hands up to answer questions. Aim to get all pupils involved. Plan for 'think, pair, share' periods for pupils to discuss your topic and self-assess when you provide the answers. All pupils should be able to answer the following two questions accurately, using specific detail. What are you doing well in this topic? And what do you need to do to improve your work?

Misconceptions

What are the main, broad misconceptions that stymie children's progress in science? Expert practitioners are highly adept at spotting and addressing the common obstacles faced by pupils (see chapter 10). This is what we mean when we talk about pedagogical content knowledge – as opposed to simply knowing in depth the material you are teaching. Try to get inside the minds of your pupils and understand the obstacles most likely to block their progress. What is the research on such misconceptions in science and how does this relate to other subjects? Can you list the main misconceptions that recur across a topic you are teaching?

SUMMARY AND KEY POINTS

- Best bets for learning relate to science teacher-pupil interactions in the classroom
- Beware alluring education myths such as learning styles
- Education research may adopt scientific approaches but doesn't have the predictive power of the natural sciences
- Evidence is necessary but never sufficient for teaching practice
- Work with peers to reflect on how evidence can help improve your practice.

Check which requirements for your initial teacher education course you have addressed through this chapter.

FURTHER RESOURCES

Elliot Major, L., & Higgins, S. (Eds.). (2019). *What works? Research and evidence for successful teaching*. Bloomsbury Education. https://www.bloomsbury.com/uk/what-works-9781472965639/

This book summarises the key research findings for a number of school approaches and presents practical steps for pupils, teachers, and school leaders. It lists some general underlying principles that apply to evidence-informed practice, including the 'Bananarama Principle' – implementing an approach is vital and just as important as its content. You might also like to listen also to the 'Podagogy' podcast broadcast by the Times Educational Supplement (Tes) (Season 7, Episode 2) https://play.acast.com/s/tes-the-education-podcast/380f1891-dac3-4a93-a3c6-9392432accf3, or watch my public lecture for the UCL Institute of Education (IOE): Apocalypse or new dawn? Social mobility and education in the post-Covid era https://www.youtube.com/watch?v=G3UZOpTWR6I&feature=emb_logo

Hattie, J. (2008). *Visible learning: A synthesis of over 800 meta-analyses relating to achievement*. Routledge.

John Hattie's book *Visible Learning* presents summaries from meta-syntheses across a range of topics, with detailed discussion on how to develop more 'visible learning' in the classroom.

Higgins, S. (2018). *Improving learning: Meta-analysis of intervention research in education*. Cambridge University Press.

This book explains the academic thinking behind meta-syntheses such as the teaching and learning toolkit. The advantages and disadvantages of this approach are explored with practical examples.

The Sutton Trust – Education Endowment Foundation Teaching and Learning Toolkit https://educationendowmentfoundation.org.uk/evidence-summaries/teaching-learning-toolkit/

The toolkit is an online freely available resource covering 30 topics, each summarised in terms of the average impact on attainment, the strength of the supporting evidence, and the cost. This is updated on a regular basis as findings from new research are added.

PUTTING RESEARCH INTO PRACTICE

Judith Bennett, Peter Fairhurst, and Alistair Moore

INTRODUCTION

The past two decades or so have seen a growing interest in many countries for education to be more *evidence-based* or *evidence-informed* – in other words, educational research should inform decisions about classroom policy and practice. A key dimension of this interest focuses on ways in which teachers can be supported to engage in *evidence-based or evidence-informed practice* and use the findings of research in their teaching. For example, the Department for Education, in its key white paper 'Educational Excellence Everywhere' (DfE, 2016b, p. 24), said that 'We want a high quality teaching profession which embraces evidence-based practice to drive up standards in schools'.

Drawing on high-quality research evidence to improve outcomes for pupils is something of a 'no-brainer', as you are very unlikely to find anyone standing up to say this is not a good idea! Science teachers want their pupils to learn effectively about key science ideas and to enjoy the science they experience in schools, and, of course, they want to make use of research that would help with these aims.

The drive for *evidence-based* or *evidence-informed* practice in education began in the second half of the 1990s. Yet, more than 20 years later, as the previous chapter shows, people are still talking about the need for it to happen. So, moving from the aspiration into classroom practice has proved elusive. Why might this be the case? Perhaps the challenge is best summarised by Paul Black and Dylan Wiliam:

> Teachers will not take up attractive-sounding ideas, albeit based on extensive research, if these are presented as general principles which leave entirely to them the task of translating them into everyday practice – their classroom lives are too busy for this to be possible. . . . What they need is a variety of living examples of implementation, by teachers with whom they

DOI: 10.4324/9781003110187-22

can identify, and from whom they can derive both the conviction and confidence that they can do better, and see concrete examples of what doing better means in practice.

(Black & Wiliam, 1998, p. 15)

What Black and Wiliam are saying is that teachers need help and support if research findings are to be 'translated' into actions and activities in schools and lessons. This chapter complements the previous one, which focuses on what works in teaching by examining what we mean by evidence-informed practice and providing some specific examples of how this can be incorporated in your science teaching. The chapter will also prove particularly useful if you want to undertake a small-scale research project focusing on your science teaching, which is often a feature or requirement of initial teacher education.

OBJECTIVES

At the end of this chapter, you should be able to:

■ Understand what people mean when they talk about *evidence-informed practice*
■ Recall the principal approaches used to gather research data and some of the key debates about what counts as 'good research'
■ Give some examples of how evidence-informed practice can help science teachers
■ Describe the key features of one large-scale project, Best Evidence Science Teaching (BEST), which helps science teachers to engage in evidence-informed practice.

WHAT DOES *EVIDENCE-BASED* OR *EVIDENCE-INFORMED* PRACTICE MEAN?

The concept of evidence-based practice originated in medicine from ideas developed in the 1970s by Cochrane (1972) to make better use of systematically gathered research evidence in the treatment of patients.

A key feature of gathering evidence in medical contexts is the randomised controlled trial (RCT) – or clinical trial – where people are randomly allocated to groups and one group gets a 'treatment' (such as a new drug) and another group gets a different treatment (often a placebo, i.e. something that resembles the new drug but does not contain the active ingredient). The outcomes for both groups are then compared. The COVID-19 pandemic has brought the use of evidence in decision-making – and the use of clinical trials – very much to the fore.

Since its introduction in medicine, the principle of evidence-based practice has since spread to other areas of decision-making, including law, public policy, management, and education. People often date the introduction of the term in education to a lecture given in 1996 by Professor David Hargreaves of the

University of Cambridge (Hargreaves, 1996). Hargreaves was very critical of educational research at the time:

> Given the huge amounts of educational research conducted over the past fifty years or more, there are few areas which have yielded a corpus of research evidence regarded as scientifically sound and as a worthwhile resource to guide professional action. . . . Just how much research is there which (i) demonstrates conclusively that if teachers change their practice from x to y there will be a significant and enduring improvement in teaching and learning, and (ii) has developed an effective method of convincing teachers of the benefits of, and means to, changing from x to y?
>
> (Hargreaves, 1996, p. 2)

He went on to argue that educational research should adopt the methods of medical research, including the use of RCTs and conducting systematic reviews of educational research literature, to assess the best available evidence.

There is debate over whether the term *evidence-based* or *evidence-informed* practice is the best to use in education. *Evidence-informed* practice appears to be more appropriate to educational settings, where research findings will be the basis of any decisions made, but these will be informed by the expertise, proficiency, and judgement that people involved in education build up through experience.

When people talk about evidence-informed practice, they often have in mind a cycle of events. This research evidence cycle is shown in Figure 21.1.

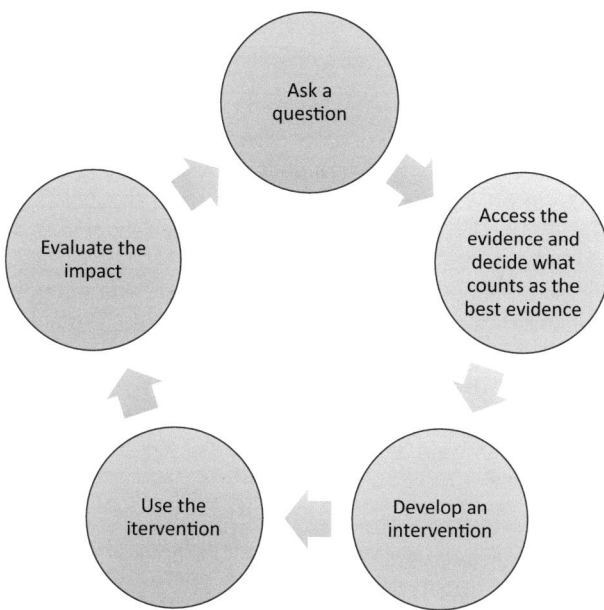

■ **Figure 21.1** A research-evidence cycle

The idea is that you use the best available research evidence to help you develop an 'intervention' (a particular approach, often meaning a strategy, type of activity or resource for teachers and pupils) to answer a particular question (e.g. what is the impact of using small-group discussions in science lessons on pupils' understanding of forces?). You then use the approach, evaluate its impact, and use what you learn from this to refine the approach.

APPROACHES USED TO GATHER RESEARCH DATA IN EDUCATION

Research evidence in education comes from carrying out a variety of different research studies, very often in educational settings such as schools. These draw on the approaches and techniques developed in the social sciences to try to answer important questions in education, such as questions about how young people learn and what makes them enjoy learning.

One helpful way of understanding the approaches used in reports of educational research studies is to consider three key features of the research: the research strategy, the research techniques, and the type of data. Box 21.1 summarises the key features of gathering evidence in education.

Box 21.1 **Key features of gathering evidence in education**

Research strategies

When people talk about *research strategy*, they are normally talking about their overall approach to the research they are conducing. Common research strategies in educational research are:

Experiments: This is probably the one with which people with a science background will be most familiar. Experiments are conducted to discover new relationships or to test out theories. In educational research, this could involve trying out a new curriculum with some pupils but not others and looking for differences in outcomes. Educational experiments can be challenging, as it may not be possible to control all the variables in a situation.

Surveys: A survey collects data from a representative – and often large – sample group to establish general trends and patterns. One example in education would be the Programme for International Student Assessment (PISA) surveys that take place every few years (www.oecd.org/pisa/).

Case studies: A case study involves looking in detail at one particular situation, for example the introduction of a new curriculum in a school.

Action research: Action research can be used to describe a small-scale study undertaken, for example, by a teacher in their school or with one or more of their classes.

Research techniques

Research techniques are the methods used to collect data. The most common techniques used in educational research are interviews, focus groups (where a group of people is given some questions or topics to discuss), questionnaires, observation (e.g. of lessons), and document studies (e.g. a new policy document, such as a national curriculum).

Types of data

The two principal types of data collected in educational research are quantitative data, which makes use of numbers (e.g. numbers of people responding in particular ways in surveys) and qualitative data, which makes use of words (e.g. what people say in interviews).

What counts as high quality research?

One of the reasons why moving towards evidence-informed practice in education has proved challenging is that people have different views over what counts as the 'best' evidence. For some people, the only evidence worth considering is evidence from RCTs, which tell you *what works*. For others, RCTs are limited, as they do not tell you why something works or, if it does not work, why it has not worked. To answer these 'why?' questions, you need to gather other sorts of evidence from, for example, surveys or talking to people. Unsurprisingly, there are people who feel the way ahead is to combine an experimental approach with other techniques in what is called a *mixed methods approach*.

In making judgements about the quality of research, the sorts of things people look at include whether or not the research has used an experimental approach, how representative the research is of normal classrooms, the size of the sample, the ways in which the data have been collected and analysed, the links between the conclusions and the data, and, for an experimental study, the *effect size*.

Effect size

The term *effect size* has become much more popular in recent years in relation to educational interventions, i.e. changes to practice, such as, for example, teaching the topic of electricity in a new way. Effect sizes are a good way of telling you how well a change is – or is not – working.

Effect size is simply a way of quantifying the size of the difference between two groups. So, you could look at the average score in a test of pupils' understanding of electricity and compare the results of a group who had experienced the new way of teaching with the results of a group taught in the more usual way and work out the effect size. If you want to know more about how effect sizes are calculated, you can look at Hattie (2012).

How evidence-informed practice can help science teachers

There are many situations where high-quality research evidence has the potential to help science teachers. Some of the research is specific to teaching science; other work is relevant to teaching more widely.

For science teachers, research evidence can help make decisions about:

■ The design of the curriculum (i.e. the sequencing of scientific ideas)
■ Approaches to teaching (such as the use of formative assessment, effective teaching of particular scientific ideas, or types of activities to use in lessons that have the potential to promote effective learning, such as the use of small group discussions)
■ Ways of engaging more of their pupils in science (such as approaches to teaching science that have the potential to improve pupil engagement, e.g. relating science to everyday life).

Table 21.1 shows examples of possible interventions that have the potential to yield evidence of use to science teachers. Some of these interventions have evidence summaries in the Education Endowment Foundation Teaching and Learning Toolkit mentioned in the previous chapter, like class size. Even when this is

■ **Table 21.1** Possible interventions that could be useful to science teachers

Aspect of teaching	Focus	Examples of questions you might ask
Organisation of teaching	Class size	Do smaller classes of 20 pupils or less lead to improved achievement for pupils? Does limiting practical work to classes of 15 pupils result improved learning of practical skills?
Approaches applicable across subjects	Assessment	What impact does formative assessment have on pupils' achievement?
Approaches used in some subjects that have potential use in science lessons	Small-group work	Do small group discussions in science lessons improve pupils' understanding of key ideas in science?
Approaches specific to science lessons	Teaching particular topics	Does a new teaching intervention lead to improved pupil understanding of particular topics, e.g. forces, electricity, evolution, the particulate nature of matter?
Approaches specific to science lessons	Relating science to everyday life	Does teaching science in the context of everyday applications increase pupils' interest and engagement in science?

the case, we would still be interested in how this evidence might inform science teaching specifically.

If you apply the research evidence cycle to one of these areas – small group discussions – it might look something like Figure 21.2:

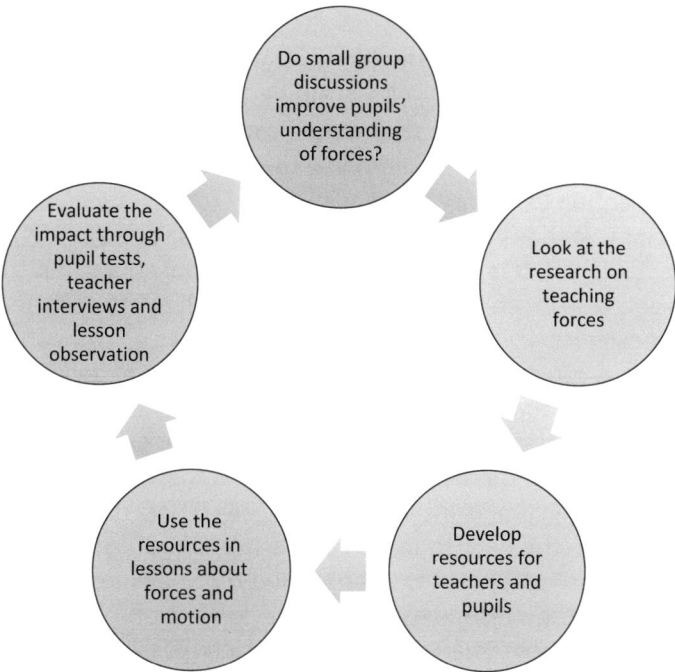

■ **Figure 21.2** A research-evidence cycle for using small-group discussions in science lessons

---

Task 21.1 **Research evidence cycle in your science lessons**

Think of something you would like to change in your science lessons. What would a research evidence cycle look like if you wanted to make your change and evaluate its effects? Plan this in line with Figure 21.1 and share your ideas with your mentor or tutor.

Implementing research findings

Before looking at some of the ways in which research evidence might help science teachers, it is interesting to think about one piece of research that is often quoted when talking about how useful research evidence is for classroom practice. This is the research about 'wait time', and it was undertaken by Mary Budd Rowe, who worked in science education. 'Wait time' is the time teachers wait for an answer after asking a class a question. Rowe's original research (Rowe, 1972, see also Rowe, 1986) showed that, when teachers asked a class a question, they typically waited less than one second for a reply – in fact, the average was 0.9s! If no one replied within that time, the teacher then rephrased the question, asked a different question, or answered the question they have asked. Rowe's research went on to demonstrate that, if teachers waited for three seconds, more pupils were likely to reply, the length and correctness of their responses increased, and, in the longer term, pupils tended to do better on tests of academic achievement.

Why might this research be so widely cited? The key is that is asks a straight-forward, focused question (how long should you wait for pupils to answer questions?), the data are easy to collect and interpret (through observing lessons), and the intervention (wait three seconds) is easy to implement. However, many educational interventions are more complex than this. If you look at some of the examples in Table 21.1, it is possible to ask focused questions, but the interventions are likely to be much more multi-faceted, and the data collected is likely to be more extensive.

For example, if you were wanting to look at the impact of an intervention consisting of a series of resources aiming to improve pupils' understanding of forces, you would have to develop the resources, design an experiment to compare the new resources with the more conventional approach, persuade schools and teachers to use them, design an instrument to collect data on understanding, and gather and analyse the data to make comparisons. You also have to make the classes you compare as similar as possible and try to ensure that teachers who are using either the new resources or the more conventional approaches teach in as similar way as possible. Then you have to consider possible outcomes, as these may influence the data you collect. What if the intervention seemed to work better in some places than others – a highly likely outcome given all the variables? Your instrument to assess pupils' understanding of forces would not tell you *why* this was the case. So, would you want to talk to teachers? Would you want to observe lessons? Deciding if and how educational interventions work is not as straightforward as it first might seem and requires considerable time and effort.

The next section of this chapter looks in detail at one intervention, The Best Evidence Science Teaching (BEST) project, which is being developed to help support science teachers engage in evidence-informed practice.

THE BEST EVIDENCE SCIENCE TEACHING (BEST) PROJECT

What is BEST?

Best Evidence Science Teaching (BEST) is a major curriculum development project that is providing a large collection of research-evidence-informed resources for science teaching at 11–16. The BEST resources are being developed by the University of York Science Education Group (UYSEG) and are being published by STEM Learning, the national centre for continuing professional development

(CPD) for science teachers. The BEST resources are all being made available to teachers free of charge to support science education, and you will find them at www.stem.org.uk/best-evidence-science-teaching. The BEST project was originally sponsored by the Salters' Institute and is now co-sponsored by the Institute of Physics. Task 21.2 introduces the BEST project and encourages you to reflect on its research problem – teaching difficult ideas in science.

Task 21.2 **Reflection on BEST introduction**

Look at the short video Introduction to *Best Evidence Science Teaching*, on the BEST homepage: www.stem.org.uk/best-evidence-science-teaching. Think about the teaching you have done so far. Have you encountered any difficulties teaching some of the more challenging ideas in science? Share your experience with other science student teachers.

The principal aims of BEST are:

■ To support teachers in engaging in evidence-informed practice (i.e. draw on research findings in their teaching)
■ To improve pupils' understanding of key concepts in science.

At the beginning of this chapter, you will have seen that people were aware of some of the problems associated with encouraging teachers to engage in evidence-informed practice. Kevan Collins noted that 'getting teachers to engage with research is far from straightforward. We need to focus our efforts on more targeted and structured approaches to disseminate evidence and support teachers' (Collins, 2016) and Paul Black and Dylan Wiliam said that teachers need 'a variety of living examples of implementation' (Black & Wiliam, 1998, p. 15). The BEST project team believes very strongly that the most effective way of supporting teachers to engage in evidence-informed practice is to work with teachers to develop resources that can be used in lessons.

The remainder of the chapter will look in more detail at the BEST resources. It also considers how the work of the project relates to the research evidence cycle.

A key question the BEST project is aiming to answer is *how can key concepts in science be taught more effectively?*

To answer this question, BEST has reviewed and used research evidence in three principal ways. These are to look at the evidence on a) the best way to sequence key concepts in science, b) how to use formative assessment most effectively in science teaching, and c) effective interventions to help develop understanding of key concepts. Where other areas of research have been relevant, these have also been used. For example, the reading age of the pupils' resources has been deliberately kept below the age of pupils to ensure pupils are able to access the information.

Developing the BEST resources

Sequencing key concepts in science

The research evidence on teaching key concepts in science led the BEST team to develop a *learning framework* for BEST structured around 15 *big ideas* in science. This learning framework underpins all the resources that have been developed for BEST. The big ideas in science are the ideas that you would want all pupils to know and understand something about when they study science at high school. Table 21.2 shows the 15 big ideas that provide the learning framework for BEST

Teaching individual topics

Each science subject has its own *key concept map*, which shows how the key concepts fit into familiar science teaching topics and how they link together to build understanding of the big ideas. Figure 21.3 shows the first section of the key concept map for Biology for ages 11–14.

You will see that each big idea is developed through a series of key concepts, organised into a number of teaching topics. These are common topics that you would expect to find on the secondary school science curriculum. For example, the big idea of the *Cellular basis of life* begins with the topic of cells, which includes four key concepts:

- Living, dead, and never been alive
- Cell size and shape
- Cell structures and their functions
- Diffusion and the cell membrane.

The second topic then goes on to look at the relationship between cells and organ systems.

You will also see that the subject map points to a possible sequence for topics across the big ideas as well as within a big idea. So, for example, it is a good idea to teach the topic of *Cells* before teaching *Changes within an organism's lifetime* and *What are health and disease?*

■ **Table 21.2** The 15 big ideas that provide the learning framework for BEST

Biology	Chemistry and earth science	Physics
The cellular basis of life Growth, reproduction, and inheritance Organisms and their environments Variation, adaptation, and evolution Health and disease	Substances and properties Particles and structure Chemical reactions Earth's atmosphere Dynamic earth	Matter Forces and motion Sound, light, and waves Electricity and magnetism Earth in space

BIOLOGY (AGE 11-14)				
BIG IDEA BCL: **THE CELLULAR BASIS OF LIFE** *Organisms are made of one or more cells, which need a supply of energy and molecules to carry out life processes.*	**BIG IDEA BHL:** **HEREDITY AND LIFE CYCLES** *Genetic information is passed from each generation to the next; this information and the environment affect the features, growth and development of organisms.*	**BIG IDEA BOE:** **ORGANISMS AND THEIR ENVIRONMENTS** *All organisms, including humans, depend on, interact with and affect the environments in which they live and other organisms that live there.*	**BIG IDEA BVE:** **VARIATION, ADAPTATION AND EVOLUTION** *Differences between organisms cause species to evolve by natural selection of better adapted individuals. The great diversity of organisms is the result of evolution.*	**BIG IDEA BHD:** **HEALTH AND DISEASE** *Organisms must stay in good health to survive and thrive; the health of an individual results from interactions between its body, behaviour, environment and other organisms.*

Topic BCL1
Cells

Key concepts:

BCL1.1 Living, dead and never been alive

BCL1.2 Cells and cell structures

BCL1.3 Cell shape and size

BCL1.4 Diffusion and the cell membrane

Topic BHL1
Inheritance and the genome

Key concepts:

BHL1.1 Heredity and genetic information

BHL1.2 The structure and function of the genome

Topic BCL2
From cells to organ systems

Key concepts:

BCL2.1 Working together – cells, tissues and organ systems

BCL2.2 Supplying cells – the human circulatory, digestive and gas exchange systems

BCL2.3 The human skeleton and muscles

Topic BVE1
Variation

Key concepts:

BVE1.1 Differences within species

BVE1.2 Changes in species over time – fossil evidence

Topic BHD1
What are health and disease?

Key concepts:

BHD1.1 Good and ill health

BHD1.2 Disease

Topic BHL2
Changes within an organism's lifetime

Key concepts:

BHL2.1 Growth
BHL2.2 Life cycles

Topic BVE2
Classification

Key concepts:

BVE2.1 Identifying and classifying organisms

Topic BHD2
Human lifestyles and health

Key concepts:

BHD2.1 Diet and exercise

■ **Figure 21.3** An extract from the key concept map for biology at ages 11–14

For each key concept, BEST provides a *progression toolkit*. This sets out a *learning focus* for the key concept and a *progression pathway*. A progression pathway is a series of sequenced learning outcomes that teachers should be able to observe as pupils' understanding of the key concept develops. Linked to each step in the pathway are: a) diagnostic questions to help provide evidence of what pupils already know and b) response activities to help teachers to challenge pupils' misunderstandings and encourage conceptual development. Figure 21.4 shows the progression toolkit for the key concept of substances.

The diagnostic questions and response activities

The diagnostic questions and response activities provided by BEST draw on the research of Paul Black and Dylan William on *formative assessment* (see Chapter 16 for more detail). The term formative assessment is used in different ways in the science education literature. In the BEST project, formative is used to describe the *function* of the assessment evidence, rather than the assessment itself. This is best summarised by Dylan Wiliam:

> Assessment functions formatively to the extent that evidence about student achievement is elicited, interpreted, and used by teachers, learners or their peers to make decisions about the next steps in instruction that are likely to be better, or better founded, than the decisions they would have made in the absence of that evidence.
>
> (Wiliam, 2011, p. 48)

Thus, the diagnostic questions in BEST are used to help teachers elicit evidence on pupils' understanding, including preconceptions and misunderstandings they may have that can form barriers to the accommodation of scientific ideas and explanations. This evidence helps teachers to decide what to do next, for example whether to move on to the next learning outcome or to spend a little more time developing pupils' understanding first.

The response activities use constructivist approaches to help move pupils' thinking towards accepted scientific understanding (see chapter 10). Often this involves challenging preconceptions and misunderstandings through *metacognition* – getting pupils to think and talk critically about what they and their peers are thinking and why. Meaning-making is fostered through activities involving, for example, small group discussion and purposeful practical work.

The Education Endowment Foundation has published a guidance report *Improving Secondary Science* (Education Endowment Foundation, 2018). This report cites BEST as a good source of diagnostic questions and of activities that promote metacognitive talk and dialogue. Task 21.3 asks you to consider what the progression pathway for the topic on forces might be, before introducing you to the associated resources for this key concept.

Learning focus	A chemical substance has a characteristic melting and boiling point and can exist in different states.			
As students' conceptual understanding progresses they can:	CONCEPTUAL PROGRESSION →			
	Recognise that a substance may exist in the solid, liquid or gas state, depending upon the temperature.	Match observations of melting (or cooling) to the temperature at which they take place.	Match observations of boiling to the temperature at which they take place.	Distinguish the scientific use of the word pure from the everyday meaning.	Distinguish a pure sample of a substance from an impure sample (mixture) by recognizing that a sharp melting point is characteristic of a pure sample of a substance.
Diagnostic questions	Possible states	Melting observations / Cooling observations	Boiling observations	Pure or mixture?	Is it pure?
Response activities	Unusual states	Comparing melting		All that glitters …	Contamination mystery

Figure 21.4 Progression toolkit: Substance

Task 21.3 **Sequencing teaching and learning for what forces do**

Without looking at Figure 21.5, see if you can arrange the sequence of the five ideas below into the correct order in which pupils should develop their ideas about 'A force makes things change the speed, direction and/or shape of an object'

■ Recognise that the motion of objects that are heavier and/or moving faster is harder to change

■ Explain changes caused by more than one force acting on an object at the same time

■ Predict correctly the changes caused by forces of difference sizes and direction on an object

■ Describe the changes, in a range of situations, which a force makes to the speed, direction and/or shape of an object.

Now compare your answer with Figure 21.5, the progression toolkit for the key concept of *What forces do*.

Figures 21.6 and 21.7 show the diagnostic question and response activity linked to the first key idea: Recognise that the motion of objects that are heavier and/or moving faster is harder to change.

BEST uses a wide variety of format in its diagnostic questions, including multiple choice answers, 'talking heads', 'cloze' (fill in the gaps) activities, sequencing an order of events, and confidence grids. These provide a rich source of information about pupils' levels of understanding. The best way to find out about these is to have a closer look at the short introductory article on BEST, available at: https://www.stem.org.uk/sites/default/files/pages/downloads/BEST_introduction_article.pdf

The teachers' notes for the BEST resources summarise the research evidence, suggest ways in which the diagnostic questions can be used, provide the correct answers to the diagnostic questions, indicate the sorts of answers you might expect from pupils, explain any misunderstandings revealed by wrong answers, suggest how to respond to misunderstandings, and suggest how the response activities can be used.

Once the BEST resources began to become available, the next step was to use them in schools – the fourth step in the research evidence cycle in Figure 21.1.

BEST in action

By 2021, much of the time spent on BEST by the project team at the University of York Science Education Group had been devoted to developing the resources. However, a key element of any intervention is gathering information on the impact of using the resources, and the BEST project team has begun a planned series of studies on *BEST in Action*. The first study (Atkinson et al.,

Learning focus	A force makes things change: the speed, direction and/or shape of an object.				
As students' conceptual understanding progresses they can:	*CONCEPTUAL PROGRESSION* →				
	Recognise that the motion of objects that are heavier and/or moving faster are harder to change. [P]	Identify situations in which a force (push or pull) is acting. [P]	Describe the changes, in a range of situations, which a force makes to the speed, direction and/or shape of an object.	Predict correctly the changes caused by forces of different sizes and direction on an object.	Explain changes caused by more than one force acting on an object at the same time. [B]
Diagnostic questions	Momentum	Is it a force?	What does this force do?	Big force, little force	An extra force
Response activities	Force or momentum?		Cycling forces		

Key:

[P] Prior understanding from earlier stages of learning

[B] Bridge to later stages of learning

Figure 21.5 Progression toolkit: What forces do

Momentum

1. Two tennis balls are thrown at you.

Ball 2

Ball 1

a. Which tennis ball is the hardest to stop?

A	Harder to stop ball 1.

B	Harder to stop ball 2.

C	Both the same.

b. How would you explain your answer to part a?

A	It is going faster.

B	It is going slower.

C	It has more force.

D	It has less force.

E	It is the same ball.

Momentum

2. You are going to push these shopping trolleys round a corner.

Trolley 2

Trolley 1

a. Which shopping trolley is harder to push round a corner?

A	Harder to turn trolley 1.

B	Harder to turn trolley 2.

C	Both the same.

b. How would you explain your answer to part a?

A	It has more force.

B	It has less force.

C	It weighs more.

D	It weighs less.

E	The wheels are the same.

■ **Figure 21.6** Diagnostic question: Momentum

Force or momentum?

Fill in the gaps to describe what happens
You should only use the words **force** and **momentum**.

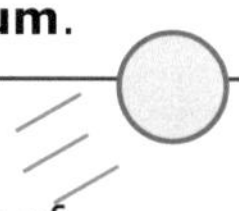

Hitting a rounders ball

John was batting. He hit the rounders ball with a lot of .... It flew quickly through the air with so much .... that James found it hard to stop. James had to use a lot of .... to stop it.

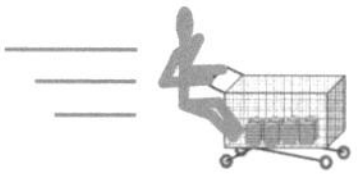

Riding on a shopping trolley

After her shopping Jane liked to run with the trolley in the car park and jump on. At the end of her shop it was heavy so she needed a lot of .... to make it move. It was hard to stop because it had a lot of .... She needed a lot of .... to stop it.

BEST™ Best Evidence Science Teaching

Developed by the University of York Science Education Group and the Salters' Institute.
This question or activity may have been edited. Download the original from **www.BestEvidenceScienceTeaching.org**

Figure 21.7 Response activity: Force or momentum

2020) explored how the BEST resources were used by teachers and the extent to which they are supporting teachers to engage in evidence-informed practice. The BEST team is gathering data on the impact of the BEST resources on pupils' understanding of key ideas, and these will be published as the analysis is completed.

The *BEST in Action Study* (Atkinson et al., 2020) was undertaken in ten case study schools in which teachers were using the BEST resources. It involved interviews with 12 teachers using the resources and observations of teachers using the resources in their lessons. Box 21.1 will remind you about what case studies are.

The key findings from the case study schools demonstrated that the BEST resources, including the research evidence summaries provided in the teacher notes included in the resources, were being used by teachers to inform how they described and explained scientific concepts, listened to pupil responses, sequenced teaching, and selected models and analogies to explain scientific ideas. The teachers' notes enabled teachers to plan lessons with increased awareness of research evidence on children's ideas in science, and the diagnostic and response items enabled teachers to gain evidence during lessons about what their pupils understood and, crucially, why. The lesson observations showed BEST resources being used consistently to identify preconceptions and as a basis for structured feedback to develop pupils' conceptual understanding. Although this initial study was small in scale, it provides what it often termed 'proof of principle'. Through 'living examples of implementation' (Black & Wiliam, 1998) and 'more targeted and structured approaches to disseminate evidence and support teachers' (Collins, 2016), the BEST resources are enabling teachers to engage in evidence-informed practice.

SUMMARY AND KEY POINTS

Now you have read this chapter, you should:

■ Understand what *evidence-informed practice* means
■ Know about how research evidence can be gathered and how research evidence fits into a research evidence cycle
■ Know about some of the key debates about what counts as 'good research'
■ Know about some of the ways in which evidence-informed practice can help science teachers
■ Be familiar with how one project, Best Evidence Science Teaching (BEST), supports science teachers to help them engage in evidence-informed practice.

Check which requirements for your initial teacher education course you have addressed through this chapter.

FURTHER RESOURCES

Bell, J., & Waters, S. (2018). *Doing your research project: A guide for first-time researchers* (7th ed.). Open University Press.

If you are thinking of designing your own small-scale research project, this book is very readable for people undertaking educational research for the first time. It describes all the basic educational research techniques and includes a number of examples.

Best Evidence Science Teaching (BEST) website. www.stem.org.uk/best-evidence-science-teaching

The BEST website contains a wide range of resources – diagnostic questions, follow-up activities, teachers' notes – for most of the topics you are likely to be teaching to pupils aged 11–14 and 14–16.

REFERENCES

Abrahams, I. (2017). Minds-on practical work for effective science learning. In K. S. Taber & B. Akpan (Eds.), *Science education. New directions in mathematics and science education* (pp. 403–413). Sense Publishers.

Abrahams, I., & Reiss, M. (2012). Practical work: Its effectiveness in primary and secondary schools in England. *Journal of Research in Science Teaching, 49*, 1035-1055.

Adesope, O. O., Trevisan, D. A., & Sundararajan, N. (2017). Rethinking the use of tests: A meta-analysis of practice testing. *Review of Educational Research, 87*(3), 659-701.

Adey, P., & Shayer, M. (2006). *Really raising standards: Cognitive intervention and academic achievement*. Routledge.

Adey, P., Shayer, M., & Yates, C. (1989). *Thinking science: The curriculum materials of the cognitive acceleration through science education (CASE) project:[Teacher's Pack]*. Nelson.

Aikenhead, G. S. (1996). Science education: Border crossing into the subculture of science. *Studies in Science Education, 27*(1), 1-52.

Akuma, F. V., & Callaghan, R. (2018). A systematic review characterizing and clarifying intrinsic teaching challenges linked to inquiry-based practical work. *Journal of Research in Science Teaching, 56*(5), 619–648.

Albalawi, S. (2021). *'Sun plants and the blue-green world': Using photo elicitation and cultural models to explore secondary students' perspectives of connecting science with the real world* [Unpublished PhD thesis, University of Manchester, Manchester].

Alexander, R. J. (2020). *A dialogic teaching companion*. Routledge. Retrieved April 27, 2021 from https://robinalexander.org.uk/dialogic-teaching/

Allen, M. (2014). *Misconceptions in primary science*. McGraw-Hill Education (UK).

Allison, S. (2017). *Making every science lesson count*. Crown House.

Anderson, K., Broderick, J. F., & Stoddard, I. (2020). A factor of two: How the mitigation plans of 'climate progressive' nations fall far short of Paris-compliant pathways. *Climate Policy, 20*(10), 1290-1304.

AQA. (2018). *GCSE combined science. Required practical handbook*. Retrieved April 23, 2020, from https://filestore.aqa.org.uk/resources/science/AQA-8464-8465-PRACTICALS-HB.PDF.

AQA. (2019). *GCSE Combined science: Trilogy (8464). Specification*. Retrieved March 7, 2021, from https://filestore.aqa.org.uk/resources/science/specifications/AQA-8464-SP-2016.PDF

AQA. (2020). *GCSE mathematics higher tier*. Retrieved Sptember 18, 2023, from https://filestore.aqa.org.uk/sample-papers-and-mark-schemes/2020/november/AQA-83001H-QP-NOV20.PDF

AQA. (2022). *Physics equations sheet: Combined science*. AQA.

REFERENCES ▪ ▪ ▪ ▪

Archer, L., Dawson, E., DeWitt, J., Seakins, A., & Wong, B. (2015a). "Science capital": A conceptual, methodological, and empirical argument for extending Bourdieusian notions of capital beyond the arts. *Journal of Research in Science Teaching, 52*(7), 922–948.

Archer, L., DeWitt, J., & Osborne, J. (2015b). Is science for us? Black students' and parents' views of science and science careers. *Science Education, 99*(2), 199–237.

Archer, L., DeWitt, J., Osborne, J., Dillon, J., Willis, B., & Wong, B. (2012). Science aspirations, capital and family habitus: How families shape children's engagement and identification with science. *American Educational Research Journal, 49*(5), 881–908.

Archer, L., Moote, J., Macleod, E., Francis, B., & DeWitt, J. (2020). *ASPIRES 2: Young people's science and career aspirations, age 10–19.* UCL Institute of Education.

Armstrong, A. C., Armstrong, D., & Spandagou, I. (2010). *Inclusive education: International policy & practice.* Sage.

Artzt, A. F., & Armour-Thomas, E. (1998). Mathematics teaching as problem solving: A framework for studying teacher metacognition underlying instructional practice in mathematics. *Instructional Science, 26*(1–2), 5–25.

Asoko, H., & Squires, A. (1998). Progression and continuity. In M. Ratcliffe (Ed.), *ASE guide to secondary science education* (pp. 175–182). Stanley Thornes.

Association for Science Education. (2020). *Safeguards in the school laboratory* (12th ed.). Association for Science Education.

Association for Science Education. (2021). *Teaching secondary biology* (3rd ed.). Hodder Education Group.

Atkinson, L., Dunlop, L., Bennett, J., Fairhurst, P., & Moore, A. (2020). Best evidence science teaching: Research evidence in action. *School Science Review, 102*(379), 55–63.

Ausubel, D. P. (1960). The use of advance organizers in the learning and retention of meaningful verbal material. *Journal of Educational Psychology, 51*(5), 267–272.

Baddeley, A. (2012). Working memory: Theories, models, and controversies. *Annual Review of Psychology, 63*, 1–29.

Beck, I., McKeown, M. G., & Kucan, L. (2013). *Bringing words to life* (2nd ed.). Guilford Press.

Beckett, L., & Wrigley, T. (2014). Overcoming stereotypes, discovering hidden capitals. *Improving Schools, 17*(3), 217–230.

Bennett, S., Maton, K., & Kervin, L. (2008). The 'digital natives' debate: A critical review of the evidence. *British Journal of Educational Technology, 39*(5), 775–786.

Berry, A. (2012). *Pedagogical content knowledge (PCK): A summary review of PCK in the context of science education research.* In J. Oversby (Ed.), *ASE guide to research in science education.* ASE.

Berry, A., & Loughran, J. (2010). *What do we know about effective CPD for developing science teachers' pedagogical content knowledge.* Paper presented at the International Seminar, Professional Reflections, National Science Learning Centre, York.

Biesta, G. A. A. (2015). What is education for? On good education, teacher judgement, and educational professionalism. *European Journal of Education, 50*(3), 75–87.

Black, P. (2017). *Assessment in science education in science education.* In K. S. Taber, & B. Akpan (Eds.), *Science education. New directions in mathematics and science education* (pp. 295–309). Sense Publishers.

Black, P. (1987). Deciding to teach. *Steam, London: ICI Science Teachers' Magazine*, No. 8.

Black, P., & Harrison, C. (2004). *Science inside the black box.* GL Assessment.

Black, P., Harrison, C., Lee, C., Marshall, B., & Wiliam, D. (2004). *Working inside the black box.* NFER Nelson.

Black, P., & Wiliam, D. (1998). *Inside the black box: Raising standards through classroom assessment.* King's College London.

Black, P., & Wiliam, D. (2018). Classroom assessment and pedagogy. *Assessment in Education: Principles, Policy and Practice, 25*(6), 551–557.

Blatchford, P. P., Bassett, P., Brown, M., Koutsoubou, C., Russell, M., Webster, R., & Rubie-Davies, C. (2009). *Deployment and impact of support staff in schools. Results from strand 2, wave 2*. Department for Children, Schools and Families.

Boaler, J. (2022). *Mathematical mindsets* (2nd ed.). Jossey-Bass.

Boohan, R. (2016). *The language of mathematics in science*. ASE.

Braund, M. (2008). *Starting science . . . again? Making progress in science learning*. SAGE.

Brock, R. (2018). Knowing is only the first step: Strategies to support the development of scientific understanding. *School Science Review, 99*(369), 116–121.

Brod, G., Werkle-Bergner, M., & Shing, Y. L. (2013). The influence of prior knowledge on memory: A developmental cognitive neuroscience perspective. *Frontiers in Behavioral Neuroscience, 7*, 139.

Brookman-Byrne, A., Mareschal, D., Tolmie, A. K., & Dumontheil, I. (2018). Inhibitory control and counterintuitive science and maths reasoning in adolescence. *pLoS ONE, 13*(6), e0198973.

Brown, S. A., & Lock, R. (2018). Enhancing intergenerational communication around climate change. In *Handbook of climate change communication: Vol. 3*. Springer.

Bruner, J. S. (1960). *The process of education*. The President and Fellows of Harvard College.

Bullough, A., & Booth, J. (2013). Science for all. *Education in Science, 251*, 12–13.

Butts, W. (1985). Children's understanding of electric current in three countries. *Research in Science Education, 15*, 127–130.

Cadwallader, S. (2018). *The impact of qualification reform on the practical skills of A level science students*. Ofqual/18/6433. Retrieved September 27, 2023, from https://assets. publishing.service.gov.uk/government/uploads/system/uploads/attachment_data/ file/747471/6433_FINAL_-_A_level_science_study_4.pdf

Calabrese Barton, A., & Upadhyay, B. (2010). Teaching and learning science for social justice: Introduction to the special issue. *Equity & Excellence in Education, 43*(1), 1–5.

Canbazoglu Bilici, S., Guzey, S. S., & Yamak, H. (2016). Assessing pre-service science teachers' technological pedagogical content knowledge (TPACK) through observations and lesson plans. *Research in Science and Technological Education, 34*(2), 237–251.

Capel, S., Leask, M., Younie, S., & Lawrence, J. (Eds.). (2022). *Learning to teach in the secondary school: A companion to school experience* (9th ed.). Routledge.

Castillo-Manzano, J. I., Castro-NuÒo, M., LÛpez-Valpuesta, L., Sanz-DÌaz, M. T., & YÒiguez, R. (2016). Measuring the effect of ARS on academic performance: A global meta-analysis. *Computers & Education, 96*, 109–121.

Caviglioli, O. (2019). *Dual coding for teachers*. John Catt Educational, Limited. Retrieved April 27, 2021, from https://www.olicav.com/

CCL (Canadian Council on Learning). (2009). *Homework helps, but not always: Lessons in Learning*. Canadian Council on Learning. Retrieved October 2, 2023, from https://files. eric.ed.gov/fulltext/ED519297.pdf

CEM (Centre for Evaluation and Monitoring). (2020). *MiDYIS standardised assessment*. Retrieved October 30, 2023, from https://www.cem.org/midyis.

Chalmers, A. F. (2013). *What is this thing called science?* (4th ed.). Open University Press.

Chi, M., Feltovich, P. J., & Glaser, R. (1981). Categorization and representation of physics problems by experts and novices. *Cognitive Science, 5*(2), 121–152.

Children Act 2004, c. 31. Retrieved September 22, 2023, from http://www.legislation.gov. uk/ukpga/2004/31/contents

Chin, C. (2007). Teacher questionning in science classrooms: Approaches that stimulate productive thinking. *Journal of Research in Science Teaching, 44*(6), 815–843.

CLEAPSS. (2017). *Health and safety induction and training for science teachers*. Retrieved September 22, 2023, from https://science.cleapss.org.uk/resource-info/g238-health-and-safety-induction-and-training-for-science-teachers.aspx

CLEAPSS. (2022). *Risk assessment. Resource SSS096a.* Retrieved September 22, 2023, from https://science.cleapss.org.uk/Resource/SSS096a-Risk-assessment.pdf

Clement, J. (1993). Using bridging analogies and anchoring intuitions to deal with students' preconceptions in physics. *Journal of Research in Science Teaching, 30*(10), 1241–1257.

Clement, J., Brown, D. E., & Zietsman, A. (1989). Not all preconceptions are misconceptions: Finding 'anchoring conceptions' for grounding instruction on students' intuitions. *International Journal of Science Education, 11*(5), 554–565.

Cochrane, A. (1972). *Effectiveness and efficiency. Random reflections on health services.* Nuffield Hospitals Trust.

Cockburn, I., Fisher, A., Mansell, E., Thind, A., & Phillips, T. (2015). *Funding for disadvantaged pupils.* National Audit Office. Retrieved October 2, 2023, from www.nao.org.uk/wp-content/uploads/2015/06/Survey-evidence-from-pupils-parents-and-school-leaders.pdf

Colburn, A. (2000). Constructivism: Science education's 'grand unifying theory'. *The Clearing House: A Journal of Educational Strategies, Issues and Ideas, 74*(1), 9–12.

Colder, J. R. (1983). In the cells of the 'Bloom Taxonomy'. *Journal of Curriculum Studies, 15*(3), 291–302.

Coll, R. K., France, B., & Taylor, I. (2005). The role of models/and analogies in science education: Implications from research. *International Journal of Science Education, 27*(2), 183–198.

Collins, K. (2016, May 20). *Support from senior leaders 'crucial' to getting teachers to engage with research.* Education Endowment Foundation (EEF) press release. https://educationendowmentfoundation.org.uk/news/support-from-senior-leaders-crucial-to-getting-teachers-to-engage-with-rese/

Collins, S., & Reiss, M. (2016). Science during primary-secondary transition. *School Science Review, 362.*

Conole, G., & Dyke, M. (2004). What are the affordances of information and communication technologies? *Research in Learning Technology, 12*(2), 113–124. Doi: 10.1080/0968776042000216183

Cooper, G., & Berry, A. (2020). Demographic predictors of senior secondary participation in biology, physics, chemistry and earth/space sciences: Students' access to cultural, social and science capital. *International Journal of Science Education, 42*(1), 151–166.

Corradi, D. M. J., Elen, J., Schraepen, B., & Clarebout, G. (2014). Understanding possibilities and limitations of abstract chemical representations for achieving conceptual understanding, *International Journal of Science Education, 36*(5), 715–734.

Cowan, N. (2010). The magical mystery four: How is working memory capacity limited, and why?. *Current Directions in Psychological Science, 19*(1), 51–57.

Craik, F., & Lockhart, R. (1972). Levels of processing: A framework for memory research. *Journal of Verbal Learning and Verbal Behavior, 11*(6), 671–684.

Crawford, C., Dearden, L., & Greaves, E. (2013). *When you are born matters: Evidence for England.* IFS Reports (No. R80). Institute for Fiscal Studies.

Criado PÈrez, C. (2019). *Invisible women: Exposing data bias in a world designed for men.* Random House.

D'Angelo, C. M., Rutstein, D., Harris, C. J., Bernard, R., & Borokhovski, E. (2014). *Simulations for STEM learning: Systematic review and meta-analysis executive summary.* SRI Education. Retrieved September 26, 2023, from https://www.sri.com/wp-content/uploads/2021/12/simulations-for-stem-learning-executive-summary.pdf

Davis, P., Florian, L., & Ainscow, M. (2004). *Teaching strategies and approaches for pupils with special educational needs: A scoping study.* DfES Publications.

Dekker, S., Lee, N., Howard-Jones, P., & Jolles, J. (2012). Neuromyths in education: Prevalence and predictors of misconceptions among teachers. *Frontiers in Psychology, 3, 429.*

Department for Education and Department of Health. (2015). *Special educational needs and disability code of practice: 0 to 25 years*. https://www.gov.uk/government/publications/send-code-of-practice-0-to-25.

Dewey, J. (1910). *How we think*. Heath.

DeWitt, J., & Archer, L. (2015). Who aspires to a science career? A comparison of survey responses from primary and secondary school students. *International Journal of Science Education, 37*(13), 2170–2192.

DeWitt, J., Osborne, J., Archer, L., Dillon, J., Willis, B., & Wong, B. (2013). Young children's aspirations in science: The unequivocal, the uncertain and the unthinkable. *International Journal of Science Education, 35*(6), 1037–1063.

Dexter, D. D., & Hughes, C. A. (2011). Graphic organizers and students with learning disabilities: A meta-analysis. *Learning Disability Quarterly, 34*(1), 51–72.

DfE (Department for Education). (2011). *Teachers' standards*. Retrieved September 22, 2023, from https://www.gov.uk/government/publications/teachers-standards

DfE (Department for Education). (2014). *GCE AS and A level subject content for biology, chemistry, physics and psychology*. Retrieved April 23, 2020, from https://www.gov.uk/government/uploads/system/uploads/attachment_data/file/593849/Science_AS_and_level_formatted.pdf

DfE (Department for Education). (2015). *National curriculum in England: Science programmes of study*. Retrieved September 22, 2023, from https://www.gov.uk/government/publications/national-curriculum-in-england-science-programmes-of-study

DfE (Department for Education). (2016a). *A guide to STEM CPD opportunities for teachers*. Retrieved October 2023, from https://assets.publishing.service.gov.uk/media/5a80237ae5274a2e87db812c/A_guide_to_STEM_CPD_opportunities_for_teachers.pdf

DfE (Department for Education). (2016b). *Educational excellence everywhere*. HMSO. Retrieved October 2023, from https://assets.publishing.service.gov.uk/government/uploads/system/uploads/attachment_data/file/508447/Educational_Excellence_Everywhere.pdf

DfE (Department for Education). (2019). *Early career framework*. Retrieved September 22, 2023, from https://www.gov.uk/government/publications/early-career-framework

DfE (Department for Education). (2022). *Health and safety: Responsibilities and duties for schools*. Retrieved September 22, 2023, from Health and safety: Responsibilities and duties for schools – GOV.UK, www.gov.uk

DfE (Department for Education). (2023). *Keeping children safe in education*. Retrieved September 22, 2023, from https://www.gov.uk/government/publications/keeping-children-safe-in-educatio–2

Didau, D., & Rose, N. (2016). *What every teacher needs to know about psychology*. John Catt.

Donnelly, J. (2001). Contested terrain or unified project? 'The nature of science' in the National Curriculum for England and Wales. *International Journal of Science Education, 23*, 181–195.

Driver, R., Guesne, E., & Tiberghien, A. (1985). Children's ideas and the learning of science. *Children's ideas in science* (pp. 1–9). Open University.

Driver, R., Leach, J., Millar, R., & Scott, P. (1996). *Young people's images of science*. Open University Press.

Driver, R., Squires, A., Rushworth, P., & Wood-Robinson, V. (2014). *Making sense of secondary science* (Routledge Education Classic ed.). Routledge.

Dunlosky, J., Rawson, K., Marsh, E., Nathan, M., & Willingham, D. (2013). Improving students' learning with effective learning techniques. *Psychological Science in the Public Interest, 14*(1), 4–58.

Earthlearningidea. (n.d.). *The toilet roll of time*. Retrieved October 30, 2023, from https://www.earthlearningidea.com/PDF/234_Toilet_roll_of_time.pdf

Easton, M. (2014). Learning the facts about learning. *BBC News*. Retrieved October 2, 2023, from www.bbc.co.uk/news/uk-30210514?sThisFB

Edmondson, K. M. (2005). Assessing science understanding through concept maps. In J. J. Mintzes, J. H. Wandersee & J. D. Novak (Eds.), *Assessing science understanding: A human constructivist view* (pp. 15–40). Elsevier Academic Press.

Education Endowment Foundation. (2018). *Improving secondary science*. Retrieved September 28, 2023, from https://educationendowmentfoundation.org.uk/education-evidence/guidance-reports/science-ks3-ks4

Education Endowment Foundation. (2019). *Improving literacy in secondary schools: Guidance report*. Retrieved October 11, 2023, from https://educationendowmentfoundation.org.uk/education-evidence/guidance-reports/literacy-ks3-ks4

Education Endowment Foundation. (2020). *Special educational needs in mainstream schools guidance report*. https://educationendowmentfoundation.org.uk/education-evidence/guidance-reports/send

El Takach, S., & Yacoubian, H. A. (2020). Science teachers' and their students' perceptions of science and scientists. *International Journal of Education in Mathematics, Science and Technology*, 8(1), 65–75.

Elliot Major, L., & Higgins, S. (2019). *What works? Research and evidence for successful teaching*. Bloomsbury Education.

Evagorou, M., & Osborne, J. (2010). The role of language in the learning and teaching of science. In J. Osborne & J. Dillon (Eds.), *Good practice in science education: What research has to say*. Open University Press.

Evans, B. (2021, April 21). Interview with beginning teacher by Caro Garrett.

Feyerabend, P. (1993). *Against method* (3rd ed.). Verso.

Findlay, M., & Bryce, T. (2012). From teaching physics to teaching children: Beginning teachers learning from pupils. *International Journal of Science Education*, 34(17), 2727–2750.

Flavell, J. H. (1976). Metacognitive aspects of problem solving. In L. B. Resnick (Ed.), *The nature of intelligence* (pp. 231–235). Lawrence Erlbaum.

Fletcher-Wood, H. (2018). *Responsive teaching*. Routledge.

Florian, L., & Black-Hawkins, K. (2011). Exploring inclusive pedagogy. *British Educational Research Journal*, 37(5), 813–828.

Fiorella, L., & Mayer, R. E. (2016). Eight ways to promote generative learning. *Educational Psychology Review*, 28(4), 717–741.

Fraillon, J., Ainley, J., Schulz, W., Friedman, T., & Duckworth, D. (2020). *Preparing for life in a digital world: IEA international computer and information literacy study 2018 international report*. Springer International Publishing.

Francis, B., Archer, L., Moote, J., DeWitt, J., MacLeod, E., & Yeomans, L. (2016). The construction of physics as a quintessentially masculine subject: Young people's perceptions of gender issues in access to physics. *Sex Roles*, 76(3), 156–174.

Francis, B., Connolly, P., Archer, L., Hodgen, J., Mazenod, A., Pepper, D., Sloan, S., Taylor, B., Tereshchenko, A., & Travers, M. C. (2017). Attainment grouping as self-fulfilling prophecy? A mixed methods exploration of self-confidence and set level among Year 7 students. *International Journal of Educational Research*, 86, 96–108.

Fraser, N. (1997). *Justice interruptus*. Routledge.

Gathercole, S. E., & Alloway, T. P. (2004). Working memory and classroom learning. *Dyslexia Review*, 15, 4–9.

Gathercole, S. E., & Alloway, T. P. (2008). *Working memory and learning: A practical guide for teachers*. Sage.

Gibbs, G. (1988). *Learning by doing: A guide to teaching and learning methods*. Further Education Unit. Oxford Polytechnic.

Gibson, J. J. (1979). *The ecological approach to visual perception*. Houghton Mifflin Har-court (HMH).

Gilbert, J. K., & Justi, R. (2016). *Modelling-based teaching in science education* (1st ed.). Springer.

Godec, S., King, H., & Archer, L. (2017). *The science capital teaching approach: Engaging students with science, promoting social justice*. University College London.

Godfrey-Smith, P. (2016). *Other minds: The octopus and the evolution of intelligent life*. William Collins.

Goldberg, R. F., & Thompson-Schill, S. L. (2009). Developmental "roots" in mature biolog-ical knowledge. *Psychological Science, 20*(4), 480–487.

Goldsworthy, A., Watson, R., & Wood-Robinson, V. (1999). *Getting to grips with graphs*. ASE.

Gonzalez, N., Moll, L., & Amanti, C. (2002). *Funds of knowledge: Theorising practices in households, communities and classrooms*. Lawrence Erlbaum Associates.

Gorski, P. C. (2016). Poverty and the ideological imperative: A call to unhook from deficit and grit ideology and to strive for structural ideology in teacher education. *Journal of Education for Teaching, 42*(4), 378–386.

Gott, R., & Duggan, S. (1995). *Investigative work in the science curriculum. Developing science and technology education*. Open University.

Graham, L. J., De Bruin, K., Lassig, C., & Spandagou, I. (2021). A scoping review of 20 years of research on differentiation: Investigating conceptualisation, characteristics, and methods used. *Review of Education, 9*(1), 161–198.

Gray, J. (2016). *Straw dogs: Thoughts on humans and other animals*. Farrar, Straus and Giroux.

Greene, B. (2020). *Until the end of time: Mind, matter, and our search for meaning in an evolving universe*. Knopf.

Grey, E., & Tall, D. (1994). Duality, ambiguity, and flexibility: A "Proceptual" view of simple arithmetic. *Journal for Research in Mathematics Education, 25*(2), 116–140.

Gurel, D. K., Eryilmaz, A., & McDermott, L. C. (2015). A review and comparison of diag-nostic instruments to identify pupils' misconceptions in science. *Eurasia Journal of Mathematics, Science and Technology Education, 11*(5), 989–1008.

Haberman, M. (2010). The pedagogy of poverty versus good teaching. *Phi Delta Kappan, 92*(2), 81–87.

Halliday, M. A. K., & Martin, J. R. (1993). *Writing science: Literacy and discursive power*. Falmer Press.

Hammer, D., & Elby, A. (2003). Tapping epistemological resources for learning physics. *The Journal of the Learning Sciences, 12*(1), 53–90.

Hammond, M. (2010). What is an affordance and can it help us understand the use of ICT in education? *Education and Information Technologies, 15*(3), 205–217.

Hammond, M. (2020). What is an ecological approach and how can it assist in under-standing ICT take-up?, *British Journal of Educational Technology, 51*(3), 853–866.

Hanushek, E. (2016). What matters for student achievement: Updating Coleman on the influence of families and schools. *Education Next, 16*, 22–30.

Harding, S. (1991). *Whose science? Whose knowledge?: Thinking from women's lives*. Cor-nell University Press.

Hardman, M. A. (2017). Models, matter and truth in doing and learning science. *School Science Review, 98*(365), 91–98.

Hargreaves, D. H. (1996). *The teacher training agency annual lecture 1996: Teaching as a research-based profession: Possibilities and prospects*. Retrieved February 26, 2024, from https://eppi.ioe.ac.uk/cms/Portals/0/PDF%20reviews%20and%20summaries/TTA%20Hargreaves%20lecture.pdf

Harlen, W. (Ed.). (2010). *Principles and big ideas of science education*. Association for Science Education.

REFERENCES ▨ ▨ ▨ ■

Harrison, A. G., & Treagust, D. F. (2000). A typology of school science models. *International Journal of Science Education, 22*(9), 1011–1026.

Hattie, J. (2012). *Visible Learning: A synthesis of meta-analyses relating to achievement* (2nd ed.). Routledge.

Hattie, J., & Yates, G. C. (2013). *Visible learning and the science of how we learn*. Routledge.

Heath, C., & Heath, D. (2007). *Made to stick: Why some ideas survive and others die*. Random House.

Hetherington, L., & Wegerif, R. (2018). Developing a material-dialogic approach to pedagogy to guide science teacher education. *Journal of Education for Teaching, 44*(1), 27–43.

Hew, K. F., Lan, M., Tang, Y., Jia, C., & Lo, C. K. (2019). Where is the "theory" within the field of educational technology research? *British Journal of Educational Technology, 50*(3), 956–971.

Higgins, S. (2018). *Improving Learning: Meta-analysis of intervention research in education*. Cambridge University Press.

Higgins, S., Katsipataki, M., Kokotsaki, D., Coleman, R., Elliot Major, L., & Coe, R. (2014). *The Sutton trust – Education endowment foundation teaching and learning toolkit*. Education Endowment Foundation.

Hillier, J. (2018). Planning for pupil understanding and learning behaviour. In I. Banner & J. Hillier (Eds.), *ASE guide to secondary science education* (pp. 89–98). ASE.

Hofer, S. I. (2015). Studying gender bias in Physics grading: The role of teaching experience and country. *International Journal of Science Education, 37*(17), 2879–2905.

Holman, J. (2017). *Good practical science*. The Gatsby Charitable Foundation. Retrieved September 27, 2023, from https://www.gatsby.org.uk/education/programmes/support-for-practical-science-in-schools

Holman, J., & Yeomans, E. (2018). *Improving secondary science: Guidance report*. Education Endowment Foundation. Retrieved September 28, 2023, from https://d2tic4wvo1iusb.cloudfront.net/production/eef-guidance-reports/science-ks3-ks4/Secondary-Science-v2.96-WEB.pdf?v=1695291250

Howard-Jones, P. (2014). Neuroscience and education: Myths and messages. *Nature Reviews Neuroscience, 15*, 817–824.

Howard-Jones, P., Ioannou, K., Bailey, R., Prior, J., Yau, S. H., & Jay, T. (2018). Applying the science of learning in the classroom. *Impact: Journal of the Chartered College of Teaching, 18*, 19.

Hughes, J., Thomas, R., & Scharber, C. (2006). Assessing technology integration: The RAT – replacement, amplification, and transformation – framework. *Society for Information Technology & Teacher Education International Conference* (pp. 1616–1620). Association for the Advancement of Computing in Education (AACE), Waynesville, NC.

Huppert, J., Lomask, S. M., & Lazarowitz, R. (2002). Computer simulations in the high school: Students' cognitive stages, science process skills and academic achievement in microbiology. *International Journal of Science Education, 24*(8), 803–821.

Husmann, P., & O'Loughlin, V. (2018). Another nail in the coffin for learning styles? Disparities among undergraduate anatomy students' study strategies, class performance, and reported VARK learning styles. *Anatomical Sciences Education, 12*(1), 6–19.

IOP (Institute of Physics). (2013). *Closing doors: Exploring gender and subject choice in schools*. Institute of Physics.

IOP (Institute of Physics). (2012). *It's different for girls: The influence of schools*. Institute of Physics.

Jenkins, E. W., & Nelson, N. W. (2005). Important but not for me: Students' attitudes towards secondary school science in England. *Research in Science and Technological Education, 23*(1), 41–57. Doi:10.1080/02635140500068435

Jeynes, W. H. (2005). A meta-analysis of the relation of parental involvement to urban elementary school student academic achievement. *Urban Education, 40*(3), 237–269.

Jones, M., Willits, J., & Dennis, S. (2015). Models of semantic memory. In J. R. Busemeyer, Z. Wang, J. T. Townsend, & A. Eidels (Eds.), *The Oxford handbook of mathematical and computational psychology* (pp. 232–254). Oxford: Oxford University Press.

Kahneman, D. (2011). *Thinking, fast and slow*. Farrar, Straus and Giroux.

Kane, M. J., & Engle, R. W. (2002). The role of prefrontal cortex in working-memory capacity, executive attention, and general fluid intelligence: An individual-differences perspective. *Psychonomic Bulletin & Review, 9*(4), 637–671.

Kennewell, S. (2001). Using affordances and constraints to evaluate the use of information and communications technology in teaching and learning. *Journal of Information Technology for Teacher Education, 10*(1-2), 101–116. Doi: 10.1080/14759390100200105

Keys, C. W., Hand, B., Prain, V., & Collins, S. (1999). Using the science writing heuristic as a tool for learning from laboratory investigations in secondary science. *Journal of Research in Science Teaching, 36*(10), 1065–1084.

Kind, V. (2004). *Beyond Appearances: Students' misconceptions about basic chemical ideas*. RSC.

Kirschner, P., Sweller, J., & Clark, R. (2006). Why minimal guidance during instruction does not work: An analysis of the failure of constructivist, discovery, problem-based, experiential, and inquiry-based teaching. *Educational Psychologist, 41*(2), 75–86.

Knight, J. (2007). *Instructional coaching: A partnership approach to improving instruction*. Corwin Press.

Knoll, A., Otani, H., Skeel, R., & Van Horn, K. (2017). Learning style, judgements of learning, and learning of verbal and visual information. *British Journal of Psychology, 108*(3), 544–563.

Kolb, D. A. (1984). *Experiential learning: Experience as the source of learning and development*. Prentice-Hall.

Kuhn, T. S. (1970). *The structure of scientific revolutions* (2nd ed.). University of Chicago Press.

Latour, B., & Woolgar, S. (1979). *Laboratory life: The social construction of scientific facts*. Sage.

Laux, K. (2018). A theoretical understanding of the literature on student voice in the science classroom. *Research in Science & Technological Education, 36*(1), 111–129.

Lederman, N., & Niess, M. (1998). 5 Apples + 4 oranges =? *School Science and Mathematics, 98*(6), 281–284.

Leinhardt, G., Zaslavsky, O., & Stein, M. (1990). Functions, graphs and graphing: Tasks, learning and teaching. *Review of Educational Research, 60*(1), 1–64.

Lemke, J. L. (2001). Articulating communities: Sociocultural perspectives on science education. *Journal of Research in Science Teaching, 38*(3), 296–316.

Lemke, J. L. (1990). *Talking science: Language, learning and values*. Ablex Publishing.

Lewandowsky, S., & Cook, J. (2020). *The conspiracy theory handbook*. http://sks.to/conspiracy

Lindner, K. T., & Schwab, S. (2020). Differentiation and individualisation in inclusive education: A systematic review and narrative synthesis. *International Journal of Inclusive Education*, 1–21.

Lowe, H., & Joffe, V. (2017). Exploring the feasibility of a classroom-based vocabulary intervention for mainstream secondary school students with language disorder. *Support for Learning, 32*(2), 110–128.

Loughran, J., Mulhall, P., & Berry, A. (2004). In search of pedagogical content knowledge in science: Developing ways of articulating and documenting professional practice. *Journal of Research in Science Teaching, 41*(4), 370–391.

Luckin, R., Bligh, B., Manches, A., Ainsworth, S., Crook, C., & Noss, R. (2012). Decoding learning: The proof, promise and potential for digital education. *Analysis and Policy Observatory NESTA*. Retrieved February 26, 2024, from https://apo.org.au/node/32254

REFERENCES ▪ ▪ ▪ ▪

Maeng, J. L., & Gonczi, A. (2019). Do plants breathe? *The Science Teacher, 86*(7), 28–34.

Malm, A. (2018). *The progress of this storm: Nature and society in a warming world*. Verso.

Mansfield, C. F., Beltman, S., Broadley, T., & Weatherby-Fell, N. (2016). Building resilience in teacher education: An evidence informed framework. *Teaching and Teacher Education, 54,* 77–87.

Marmot, M. (2020). Health equity in England: The Marmot review 10 years on. *BMJ, 368*.

Martin, K., & Miller, E. (1988). Storytelling and science. *Language Arts, 65*(3), 255–259.

Matthews, M. R. (1993). Constructivism and science education: Some epistemological problems. *Journal of Science Education and Technology, 2*(1), 359–370.

Mayer, R. E. (2002). Cognitive theory and the design of multimedia instruction: An example of the two-way street between cognition and instruction. *New Directions for Teaching and Learning, 2002*(89), 55–71.

Mayer, R. E. (2009). Constructivism as a theory of learning versus constructivism as a prescription for instruction. In S. Tobias & T. M. Duffy (Eds.), *Constructivist instruction success or failure?* (pp. 184–200). Routledge/Taylor & Francis Group.

Mehan, H. (1979). *Learning lessons: Social organisation in the classroom*. Harvard University Press.

Mercer, N., Dawes, L., Wegerif, R., & Sams, C. (2004). Reasoning as a scientist: Ways of helping children to use language to learn science. *British Educational Research Journal, 30*(3), 359–377.

Merton, R. K. (1973). *The sociology of science: Theoretical and empirical investigations*. University of Chicago Press.

Messiou, K., & Hope, M. A. (2015). The danger of subverting students' views in schools. *International Journal of Inclusive Education, 19*(10), 1009–1021.

Miles, S., & Howes, A. (Eds.). (2015). *Photography in educational research: Critical reflections from diverse contexts*. Routledge.

Millar, R. (1988). Teaching physics as a non-specialist: The in-service training of science teachers. *Journal of Education for Teaching, 14*(1), 39–53.

Millar, R. (2009). *Analysing practical activities to assess and improve effectiveness: The Practical Activity Analysis Inventory (PAAI)*. Centre for Innovation and Research in Science Education, University of York. Retrieved April 23, 2020, from http://www.york.ac.uk/depts/educ/research/ResearchPaperSeries/index.htm

Millar, R. (2014). Teaching about energy: From everyday to scientific understandings. *School Science Review, 96*(354), 45–50.

Miller, D. I., Nolla, K. M., Eagly, A. H., & Uttal, D. H. (2018). The development of children's gender-science stereotypes: A meta-analysis of 5 decades of US draw-a-scientist studies. *Child Development, 89*(6), 1943–1955.

Mishra, P., & Koehler, M. (2006). Technological pedagogical content knowledge: A framework for teacher knowledge. *The Teachers College Record, 108*(6), 1017–1054.

Moore, D., & Black, A. (2021). Supporting beginning teachers to work with pupils with special educational needs and disability. In S. Salahjee (Ed.), *Mentoring science teachers in the secondary school* (pp. 291–304). Routledge.

Mortimer, E., & Scott, P. (2003). *Mearning making in secondary science classrooms*. Open University Press.

Mostafa, T., Echazarra, A., & Guillou, H. (2018). *The science of teaching science: An exploration of science teaching practices in PISA 2015*. OECD Education Working Paper No. 188.

Muijs, D., & Reynolds, D. (2011). *Effective teaching, evidence and practice* (3rd ed.). Sage.

Mujtaba, T., & Reiss, M. J. (2013). Inequality in experiences of physics education: Secondary school girls' and boys' perceptions of their physics education and intentions to continue with Physics after the age of 16. *International Journal of Science Education, 35*(11), 1824–1845.

Murphy, P., & Whitelegg, E. (2016). Girls and physics: Continuing barriers to 'belonging'. *The Curriculum Journal, 17*(3), 281–305.

Murray, I., & Reiss, M. (2005). The student review of the science curriculum. *School Science Review, 87*(318), 83–93.

NASA. (2003). *Solar system sizes*. Retrieved August 2023, from https://solarsystem.nasa.gov/system/downloadable_items/968_solarsys_scale.jpg

Naylor, S., & Keogh, B. (2000). *Science concept cartoons*. Millgate House.

Newton, P. M., da Silva, A., & Berry, S. (2020). The case for pragmatic evidence-based Higher Education: A useful way forward. *Frontiers in Education, 5*, 271.

NGSS (Next Generation Science Standards). (2023). *Read the standards*. Retrieved September 28, 2023, from http://www.nextgenscience.org/search-standards

NHS Ayrshire and Arran. (2019). *Understanding Fetal Alcohol Spectrum Disorder (FASD): What educators need to know*. https://www.nhsaaa.net/media/8391/fasd_whateducatorsneedtoknow.pdf

Norwich, B. (2010). Dilemmas of difference, curriculum and disability: International perspectives. *Comparative Education, 46*(2), 113–135.

Nott, M., & Wellington, J. (1993). Your nature of science profile: An activity for science teachers. *School Science Review, 75*(270), 109–112.

NRICH. (2011). *Creating a low threshold high ceiling classroom*. Retrieved October 2, 2023, from https://nrich.maths.org/7701

NRICH. (2023a). *Fill me up*. Retrieved September 28, 2023, from https://nrich.maths.org/7419

NRICH. (2023b). *Mixing lemonade*. Retrieved September 28, 2023, from https://nrich.maths.org/6870

Nuthall, G. (2007). *The hidden lives of learners*. NZCER Press.

Nyachwaya, J. M., & Gillaspie, M. (2016). Features of representations in general chemistry textbooks: A peek through the lens of cognitive load theory. *Chemistry Education Research and Practice, 17*, 58–71.

Ofqual (Office of Qualifications and Examinations Regulation). (2015a). *Assessment of practical work in new science GCSEs – Summary*. Ofqual. Retrieved April 23, 2020, from https://assets.publishing.service.gov.uk/government/uploads/system/uploads/attachment_data/file/408513/2015-03-03-assessment-of-practical-work-in-new-science-gcses-summary.pdf.

Ofqual (Office of Qualifications and Examinations Regulation). (2015b). *GCSE subject level conditions and requirements for science*. Ofqual. Retrieved September 28, 2023, from https://assets.publishing.service.gov.uk/government/uploads/system/uploads/attachment_data/file/819655/gcse-subject-level-conditions-and-requirements-for-combined-science.pdf

Ofsted (Office for Standards in Education, Children's Services and Skills). (2019). *Education inspection framework*. Retrieved April 13, 2021 from https://www.gov.uk/government/publications/education-inspection-framework

Ofsted. (2015). A-*level subject take-up: Numbers and proportions of girls and boys studying A-level subjects in England*. Retrieved June 16, 2020, from https://www.gov.uk/government/publications/a-level-subject-take-up

Ogborn, J., Kress, G., Martins, I., & McGillicuddy, K. (1996). *Explaining science in the classroom*. Open University Press.

Oliver, M. (2005). The problem with affordance. *E-Learning and Digital Media, 2*(4), 402–413.

Oliver, M. (2013). Learning technology: Theorising the tools we study. *British Journal of Educational Technology, 44*(1), 31–43.

Opposs, D. (2016). Whatever happened to school-based assessment in England's GCSEs and a levels? *Perspectives in Education, 34*(4), 52–61.

Osborne, J. (2014). Teaching scientific practices: Meeting the challenge of change. *Journal of Science Teacher Education, 25*, 177-196.

Osborne, J., & Dillon, J. (2008). *Science education in Europe*. Nuffield Foundation.

Osborne, J., & Dillon, J. (2010). *Good practice in science teaching: What research has to say* (2nd ed.). Open University Press.

Osborne, J. F., Henderson, J. B., MacPherson, A., Szu, E., Wild, A., & Yao, S. Y. (2016). The development and validation of a learning progression for argumentation in science. *Journal of Research in Science Teaching, 53*(6), 821-846.

Osborne, R. (1983). Towards modifying children's ideas about electric current. *Research in Science & Technological Education, 1*(1), 73-82.

Osborne, R. (2015). *An ecological approach to educational technology: Affordance as a design tool for aligning pedagogy and technology* [PhD thesis]. http://hdl.handle.net/10871/16637

Osborne, R. (2019). Rethinking education technology from digital tool to digital place: New perspectives, new affordances. *Impact*, 20-22.

Ozdemir, G. (2017). Utilizing concrete manipulatives in contextually distinct situations to assess middle school students' meanings of force. *International Journal of Education in Mathematics, Science and Technology, 5*(3), 187-202.

Pais, A. (1986). *Inward bound: Of matter and forces in the physical world*. Oxford University Press.

Paivio, A. (1971). *Imagery and verbal processes*. Holt, Rinehart, and Winston.

Paremoer, L., Nandi, S., Serag, H., & Baum, F. (2021). Covid-19 pandemic and the social determinants of health. *BMJ, 372*.

Paul, R., & Elder, L. (2007). Critical thinking: The art of Socratic questioning. *Journal of Developmental Education, 31*(1), 36.

Paul, R., & Elder, L. (2019). *The thinker's guide to Socratic questioning*. Rowman & Littlefield.

Piaget, J. (1929). *The child's conception of the world* (J. Tomlinson & A. Tomlinson, Trans.). Routledge & Kegan Paul Ltd.

Piaget, J. (1977). *The development of thought: Equilibrium of cognitive structures*. Viking.

Pickett, K., & Wilkinson, R. (2010). *The spirit level: Why equality is better for everyone*. Penguin.

Pines, A. L., & West, L. H. T. (1986). Conceptual understanding and science learning: An interpretation of research within a sources-of-knowledge framework. *Science Education, 70*(5), 583-604.

Pinker, S. (2002). *The blank slate: The modern denial of human nature*. Penguin.

Popper, K. R. (1934/1972). *The logic of scientific discovery*. Hutchinson.

Prensky, M. (2001). Digital natives, digital immigrants Part 1. *On The Horizon - The Strategic Planning Resource for Education Professionals, 9*(5), 1-6.

Pring, R. (1971). Bloom's taxonomy: A philosophical critique (2), *Cambridge Journal of Education, 1*(2), 83-91.

Puentedura, R. (2006). *Transformation, technology, and education* [Blog post]. Retrieved February 26, 2024, from http://hippasus.com/resources/tte/

Pugh, K. R., Mencl, W. E., Jenner, A. R., Katz, L., Frost, S. J., Lee, J. R., Shaywitz, S. E., & Shaywitz, B. A. (2000). Functional neuroimaging studies of reading and reading disability (developmental dyslexia). *Mental Retardation and Developmental Disabilities Research Reviews, 6*(3), 207-213.

Qualter, A., Strang, J., Swatton, P., & Taylor, R. (1990). *Exploration: A way of learning science*. Blackwell.

Quigley, A., & Coleman, R. (2019). *Improving literacy in secondary schools: Guidance report*. Education Endowment Foundation. Retrieved September 28, 2023, https://d2tic4wvo1iusb.cloudfront.net/production/eef-guidance-reports/literacy-ks3-ks4/EEF_KS3_KS4_LITERACY_GUIDANCE.pdf?v=1695883083

Quigley, A., Muijs, D., & Stringer, E. (2018). *Metacognition and self-regulated learning: Guidance report*. Education Endowment Foundation. Retrieved September 28, 2023, from https://d2tic4wvo1iusb.cloudfront.net/eef-guidance-reports/metacognition/EEF_Metacognition_and_self-regulated_learning.pdf?v=1676504749

Raved, L., & Zvi Assaraf, O. B. (2011). Attitudes towards science learning among 10th-grade students: A qualitative look. *International Journal of Science Education, 33*(9), 1219–1243.

Redish, E. F. (2017). Analysing the competency of mathematical modelling in physics. In T. Greczylo & E. Debowska (Eds.), *Key competences in physics teaching and learing* (pp. 25–40). Springer.

Redish, E. F., & Kuo, E. (2015). Language of physics, language of math: Disciplinary culture and dynamic epistemology. *Science and Education, 24*, 561–590.

Reiss, M. J. (1993). *Science education for a pluralist society*. Open University Press.

Reiss, M. J. (2003). Science education for social justice. *Social Justice, Education and Identity*, 153–165.

Reiss, M. J., Abrahams, I., & Sharpe, R. (2012). *Improving the assessment of practical work in school science*. The Gatsby Foundation. Retrieved March 29, 2024, https://www.gatsby.org.uk/uploads/education/reports/pdf/improving-the-assessment-of-practical-work-in-school-science.pdf

Rivet, A. E., & Kastens, K. A. (2012). Developing a construct-based assessment to examine students' analogical reasoning around physical models in Earth Science. *Journal of Research in Science Teaching, 49*(6), 713–743.

Rizzo, K. L., & Taylor, J. C. (2016). Effects of inquiry-based instruction on science achievement for students with disabilities: An analysis of the literature. *Journal of Science Education for Students with Disabilities, 19*(1), 2.

Roediger, H. L., & Karpicke, J. D. (2006). Test-enhanced learning. *Psychological Science, 17*(3), 249–255.

Roediger, H. L., & Pyc, M. A. (2012). Inexpensive techniques to improve education: Applying cognitive psychology to enhance educational practice. *Journal of Applied Research in Memory and Cognition, 1*(4), 242–248.

Rogers, B. (2018). *The big ideas in physics and how to teach them: Teaching physics 11–18*. Routledge.

Rosenshine, B. (2012). Principles of instruction: Research-based strategies that all teachers should know. *American Educator, 36*(1), 12–19.

Rouse, M. (2008). Developing inclusive practice: A role for teachers and teacher education? *Education in the North, 16*, 1–20.

Rousell, D., & Cutter-Mackenzie-Knowles, A. (2019). A systematic review of climate change education: Giving young people a 'voice' and a 'hand' in redressing climate change. *Children's Geographies, 18*(2), 191–208.

Rowe, M. B. (1972). *Wait-time and rewards as instructional variables: Their influence on language, logic and fate control*. Paper presented at the annual meeting of the National Association for Research on Science Teaching, Chicago.

Rowe, M. B. (1986). Wait time: Slowing down may be a way of speeding up. *Journal of Teacher Education, 37*(1), 43–50.

Rowlands, S., Graham, T., Berry, J., & Mcwilliam, P. (2007). Conceptual change through the lens of Newtonian mechanics. *Science & Education, 16*(1), 21–42.

Royal Society of Chemistry. (2022). *Practical resources*. Retrieved January 24, 2022, from https://edu.rsc.org/resources/practical.

Rudduck, J., & Fielding, M. (2006). Student voice and the perils of popularity. *Educational Review, 58*(2), 219–231.

Scantlebury, K., Tai, T., & Rahm, J. (2007). "That don't look like me." Stereotypic images of science: Where do they come from and what can we do with them? *Cultural Studies of Science Education, 1*(3), 545–558.

Schipper, T. M., van der Lans, R. M., de Vries, S., Goei, S. L., & van Veen, K. (2020). Becoming a more adaptive teacher through collaborating in Lesson Study? Examining the influence of Lesson Study on teachers' adaptive teaching practices in mainstream secondary education. *Teaching and Teacher Education, 88*, 102961.

Schön, D. A. (1983). *The reflective practitioner: How professionals think in action.* Temple Smith.

Schwann, T. (1839). *Mikroskopische Untersuchungen über die Uebereinstimmung in der Struktur und dem Wachsthum der Thiere und Pflanzen* (Microscopic investigations on the similarity of structure and growth of animals and plants).

Schwarz, C. V., Reiser, B. J., Davis, E. A., Kenyon, L., AchÈr, A., Fortus, D., Shwartz, Y., Hug, B., & Krajcik, J. (2009). Developing a learning progression for scientific modeling: Making scientific modeling accessible and meaningful for learners. *Journal of Research in Science Teaching* (46)6, 632–654.

Science Community Representing Education (SCORE). (2008). *Benchmark tools for practical work in science.* Retrieved January 24, 2022, from https://www.iop.org/education/support-school-college-physics-teachers/benchmark-tools#gref

Science and Plants for Schools. (2022). *A-level set practicals – Microscopy of root tip mitosis.* Retrieved January 24, 2022, from https://www.saps.org.uk/secondary/teaching-resources/1358-a-level-set-practicals-microscopy-of-root-tip-mitosis

Scott, P. (1987). *A constructivist view of learning and teaching in science.* Children's Learning in Science Project. The University of Leeds. Retrieved October 30, 2023, from https://www.stem.org.uk/elibrary/collection/3069

Secondary National Strategy. (2009). *Developing writing in science.* Retrieved April 27, 2021, from https://www.stem.org.uk/elibrary/resource/31727

Sharpe, R., & Abrahams, I. (2020). Secondary school students' attitudes to practical work in biology, chemistry and physics in England. *Research in Science & Technological Education, 38*(1), 84–104.

Sharples, J., Webster, R., & Blatchford, P. (2018). *Making best use of teaching assistants: Guidance report.* EEF. Retrieved October 2, 2023, from https://educationendowmentfoundation.org.uk/education-evidence/guidance-reports/teaching-assistants

Shayer, M., & Adey, P. (1981). *Towards a science of science teaching: Cognitive development and curriculum demand.* Heinemann Educational Books.

Shulman, L. S. (1986). Those who understand: Knowledge growth in teaching. *Educational Researcher, 15*(2), 4–14.

Shtulman, A. (2017). *Scienceblind: Why our intuitive theories about the world are so often wrong.* Basic Books.

Skinner, E. A., Kindermann, T. A., & Furrer, C. J. (2009). A motivational perspective on engagement and disaffection: Conceptualization and assessment of children's behavioral and emotional participation in academic activities in the classroom. *Educational and Psychological Measurement, 69*(3), 493–525.

Smale-Jacobse, A. E., Meijer, A., Helms-Lorenz, M., & Maulana, R. (2019). Differentiated instruction in secondary education: A systematic review of research evidence. *Frontiers in Psychology, 2366.*

SQA. (2022). *National 4 science.* Retrieved January 24, 2022, from https://www.sqa.org.uk/sqa/47426.html

Starling, A., Lo, Y. Y., & Rivera, C. J. (2015). Improving science scores of middle school students with learning disabilities through engineering problem solving activities. *Journal of the American Academy of Special Education Professionals, 98*, 113.

STEM Library. (2020a). *Children's learning in science project.* Retrieved October 30, 2023, from https://www.stem.org.uk/elibrary/collection/3069

STEM Library. (2020b). *QCA schemes of work for science.* Retrieved October 30, 2023, from https://www.stem.org.uk/elibrary/resource/29139

Sutton, C. (1992). *Words, science, and learning: Developing science and technology education*. Open University Press.

Sweller, J. (2009). What human cognitive architecture tells us about constructivism. In S. Tobais & T. M. Duffy (Eds.), *Constructivist instruction: Success or failure?* (pp. 127–143). Routledge & Kegan Paul Ltd.

Sweller, J. (2011). Cognitive load theory. In J. P. Mestre, & H. B. Ross (Eds.), *Psychology of Learning and Motivation*: Cognition in Education, 55, 37–76. Elsevier.

Sweller, J., Van Merrienboer, J. J. G., & Paas, F. (2019). Cognitive architecture and instructional design: 20 years later. *Educational Psychology Review, 31*(2), 261–292.

Szyjka, S. P., & Mumba, F. (2009). Preparing science teachers for inclusive classrooms: Components of a suggested model for science teacher training. *Southeastern Teacher Education Journal, 2*(4).

Taber, K. S. (2009). *Progressing science education: Constructing the scientific research programme into the contingent nature of learning science*. Springer.

Taber, K. S. (2013). Revisiting the chemistry triplet: Drawing upon the nature of chemical knowledge and the psychology of learning to inform chemistry education. *Chemistry Education Research and Practice, 14*, 156–158.

Taber, K. S. (2017). Models and modelling in science and science education. In K. S. Taber & B. Akban (Eds.), *Science education* (pp. 263–278). Brill.

Tawney, D. (1981). *Accidents in school laboratories: A report of an investigation*. ASE. Retrieved September 22, 2023, from https://www.ase.org.uk/sites/default/files/General%20health%20and%20safety%20PDFs/PDFs/Accidents%20in%20school%20laboratories.pdf

Thadani, V., Cook, M. S., Griffis, K., Wise, J. A., & Blakey, A. (2010). The possibilities and limitations of curriculum-based science inquiry interventions for challenging the "pedagogy of poverty". *Equity & Excellence in Education, 43*(1), 21–37.

Toplis, R. (2015). Practical work. In R. Toplis (Ed.), *Learning to teach science in the secondary school* (4th ed.). Routledge.

Turşucu, S., Spandaw, J., Flipse, S., & de Vries, M. J. (2017). Teachers' beliefs about improving transfer of algebraic skills from mathematics into physics in senior pre-university education. *International Journal of Science Education, 39*(5), 587–604.

Van Driel, J. H., & Verloop, N. (1999). Teachers' knowledge of models and modelling in science. *International Journal of Science Education, 21*(11), 1141–1153.

van Geel, M., Keuning, T., FrËrejean, J., Dolmans, D., van Merriînboer, J., & Visscher, A. J. (2019). Capturing the complexity of differentiated instruction. *School Effectiveness and School Improvement, 30*(1), 51–67.

Villanueva, M. G., & Hand, B. (2011). Science for all: Engaging students with special needs in and about science. *Learning Disabilities Research & Practice, 26*(4), 233–240.

Vosniadou, S. (2012). Reframing the classical approach to conceptual change: Preconceptions, misconceptions and synthetic models. In *Second international handbook of science education* (pp. 119–130). Springer.

Vygotsky, L. S. (1962). *Thought and language*. Martino Publishing.

Vygotsky, L. S. (1931). The genesis of higher mental functions. In J. V. Wertsch (Ed.), *The concept of activity in Soviet psychology* (pp. 144–188). Sharpe.

Walsh, H. (2019). How to reflect on your use of models. *Education in Chemistry*. Retrieved September 22, 2023, from https://edu.rsc.org/feature/reflect-on-your-use-of-models/3010509.article

Warren, D. (2019). Supporting transition: Moving smoothly from primary to secondary science. *STEM Learning Magazine. Secondary*. 12. Retrieved March 29, 2024, from https://magazines.stem.org.uk/stem-learning-magazine-secondary-12.html?_ga=2.142397942.517315859.1711713875-1774393363.1708355444

Watkins, C., Carnell, E., & Lodge, C. (2007). *Effective learning in classrooms*. Sage.

REFERENCES ▪ ▪ ▪ ▪

Watson, J. (1968). *The double Helix: A personal account of the discovery of the structure of DNA*. Weidenfeld and Nicholson.

Watts, M. (2014). Explanations and explaining in science: You can't force people to understand. In M. Watts (Ed.), *Debates in science education* (pp. 132–144). Routledge.

Webster, R., & Blatchford, P. (2019). Making sense of 'teaching', 'support' and 'differentiation': The educational experiences of pupils with education, health and care plans and statements in mainstream secondary schools. *European Journal of Special Needs Education, 34*(1), 98–113.

Webster, R., Russell, A., & Blatchford, P. (2015). *Maximising the impact of teaching assistants*. Routledge.

Weinstein, Y., Madan, C. R., & Sumeracki, M. A. (2018). Teaching the science of learning. *Cognitive Research: Principles and Implications, 3*(2).

Wellington, J. (2005). Has ICT come of age? Recurring debates on the role of ICT in education, 1982–2004. *Research in Science & Technological Education, 23*(1), 25–39.

Wellington, J., & Osborne, J. (2001). *Language and literacy in science education*. Open University Press.

Wiliam, D. (2011). *Embedded formative assessment*. Solution Tree.

Willingham, D. (2009). *Why don't students like school?* Jossey Bass.

Wirebring, L. K., Wiklund-Hörnqvist Eriksson, J., Andersson, M., Jonsson, B., & Nyberg, L. (2015). Lesser neural pattern similarity across repeated tests is associated with better long-term memory retention. *Journal of Neuroscience, 35*(26), 9595–9602.

Wisniewski, B., Zierer, K., & Hattie, J. (2020). The power of feedback revisited: A meta-analysis of educational feedback research. *Frontiers in Psychology, 10*(3087).

Wong, V. (2018). *The relationship between school science and mathematics education*. King's College London.

Wong, V., & Dillon, J. (2019). 'Voodoo maths', asymmetric dependency and maths blame: Why collaboration between school science and mathematics teachers is so rare. *International Journal of Science Education, 41*(6), 782–802.

Wood, D., Bruner, J. S., & Ross, G. (1976). The role of tutoring in problem solving. *Journal of Child Psychology and Psychiatry, and Allied Disciplines, 17*(2), 89–100.

Zacharia, Z. C., & de Jong, T. (2014). The effects on students' conceptual understanding of electric circuits of introducing virtual manipulatives within a physical manipulatives-oriented curriculum. *Cognition and Instruction, 32*(2), 101–158.

INDEX

Note: Page numbers in *italics* indicate a figure and page numbers in **bold** indicate a table on the corresponding page. Page numbers followed by *"b"* refer to boxes/tasks.

INDEX